ECOLOGICAL RESTORATION IN THE MIDWEST

A BUR OAK BOOK

Holly Carver, series editor

ECOLOGICAL

RESTORATION

IN THE

MIDWEST

PAST, PRESENT,

AND FUTURE

Edited by Christian Lenhart

and Peter C. Smiley Jr.

UNIVERSITY OF IOWA PRESS, IOWA CITY

University of Iowa Press, Iowa City 52242
Copyright © 2018 by the University of Iowa Press
www.uipress.uiowa.edu
Printed in the United States of America

Design by Ashley Muehlbauer

The University of Iowa Press is a member of Green Press
Initiative and is committed to preserving natural resources.

ISBN 978-1-60938-573-6 (pbk)
ISBN 978-1-60938-574-3 (ebk)

Printed on acid-free paper

Cataloging-in-Publication data is on file
with the Library of Congress.

CONTENTS

The midwestern United States is known around the world as the cradle of ecological restoration. Pioneering projects to remake prairies in the 1930s at the University of Wisconsin–Madison Arboretum became a movement that gradually gained momentum, fueled by the newly emerging scientific discipline of ecology, by the success of these early projects, and—most importantly—by society's embrace of a conservation ethic. With Aldo Leopold as eloquent messenger and dedicated boots-on-the-ground conservationist, land stewardship became an imperative for many across the region and beyond. The rapid transformation of the Midwest's landscape—breaking prairie, felling forests, straightening streams, and draining wetlands—had reached an apogee only a few decades before these first attempts at restoration. Being able to reverse even a little of that habitat loss undoubtedly was then—as it is today—an emotionally satisfying enterprise.

Our attempts to restore a wide array of midwestern ecosystems over the past eighty-five years have taught us a great deal. We take on projects that are increasingly larger and more technically challenging. Yet we are more acutely aware than ever of the constraints and costs of reversing human-caused changes to nature. Reading this book is an opportunity to immerse yourself in the story of the restoration practice in the region—told as a story of people striving to do more, more wisely, applying foundational concepts that have shaped the practice in the Midwest.

The focus of this book is on how we have advanced restoration practice in the Midwest rather than attempting to be a compendium of techniques. This is wise, as the book illustrates, because most restorations of significant scope and scale are collaborative efforts involving a team of people with complementary skills and knowledge. For example, the stream restoration projects described in the book required teams with collective expertise in civil engineering, stream ecology, fisheries biology, soil, and vegetation. Urban projects involve planners and landscape architects, as well as experts who can apply their geotechnical and ecological knowledge. Because the book's authors share with us the deep historical roots of midwestern practices and projects, we also know that restoration has long been integrative and multidisciplinary. The many detailed cases in this book are also valuable because they are not stories of perfection but of successes tinged with setbacks that people worked to overcome. The cases featured in the book describe restoration projects where people made a commitment to a place—and stuck with it, even through major challenges. This book shows what has been required and continues to be needed to move restoration practice forward.

The editors of the book, Chris Lenhart and Peter "Rocky" Smiley Jr., assembled a team of experts who share their insights on lessons learned from a diverse array of ecological restorations, both terrestrial and aquatic and in both urban and agricultural settings. If you are new to ecological restoration in the Midwest, reading this book will orient you to the tremendous variety of projects going on today. If you are an experienced restorationist, chances are you are much more familiar with practices that relate to your specialty—typically either terrestrial or aquatic. This book is unusual in its balance and a great way to learn about a broad variety of projects.

Are the approaches used today to restore midwestern ecosystems relevant elsewhere, or is the region's importance to the field mostly historical? Or, to state it more plainly, will this book be useful to restorationists based in parts of the world other than the Midwest? Certainly, ecosystem restoration in each region of the world faces different constraints and has different opportunities—but it's important not to be too distracted by these differences. Many of the most important stressors degrading our ecosystems are global problems (and so, common challenges)—climate change, invasive species, land degradation, pollution. And, as in the midwestern United States, many restorations are being pursued in highly transformed landscapes, with legacies related to past land use that cannot be erased as part of restoration. There are many insights to be gained from an awareness of how practitioners think about and respond to the challenges they encounter. This book provides a rich level of detail about dozens of projects, which I believe will be broadly valuable.

The midwestern United States does have some particular advantages for ecological restoration. What comes to mind for many, I suspect, are the region's environmental conditions that are amenable to ecological recovery—soils that are generally rich and growing seasons that aren't too harsh. Let me suggest that even greater advantages are the well-developed communities of ecological restoration practice that have existed now for decades in many parts of the region, a legacy of Aldo Leopold and others' call for a land ethic. These social-professional networks connect innovators of native seed harvesting and planting equipment to seed vendors and native plant nurseries to consultancies and nonprofit service organizations to landowners to policy makers and to agencies that implement restoration programs, and so on. Words in a sentence can't easily portray the complexity of these networks, complexity that perhaps, as with ecosystems, conveys stability and resilience. This may be why the midwestern United States continues to be an incubator of innovation for ecological restoration, as the many examples in this book highlight. This book is a great way for anyone interested in restoration—anywhere in the world—to be inspired and better equipped to take on new challenges.

SUSAN GALATOWITSCH
Professor and Department Head, Fisheries,
Wildlife and Conservation Biology
University of Minnesota
Author, *Ecological Restoration* and *Restoring Prairie Wetlands*

In 2010, Chris Lenhart proposed the idea of writing a book devoted to ecological restoration in the Midwest to the Board of Directors of the Midwest–Great Lakes Chapter of the Society for Ecological Restoration during one of its monthly conference calls. The response from the board was highly enthusiastic, and many volunteered to help with the writing. Peter Smiley Jr., who at the time was the chapter president, then proposed that the chapter form a committee to support the development of the book. Chris later recruited Peter to serve as coeditor, and then the two of them sought out regional experts for each of the chapters. The contributing authors were provided the objective for the entire book and a general objective for each chapter. The authors then had the freedom to individualize the chapters to reflect their experiences and expertise. Each chapter underwent two reviews by the editors and one review by two anonymous reviewers selected by the University of Iowa Press.

We extend our gratitude to the fifteen contributing authors for their dedication to this project, as their contributions resulted in a novel compilation of ecological restoration concepts and case studies from across the Midwest. We thank the two anonymous reviewers for their detailed and thoughtful reviews of the first draft of this book. We also thank Chad Bladlow, Meredith Cornett, Emily Deering, Natasha DeVoe, Susan Galatowitsch, Brad Gordon, Jennifer M. Grieser, Evelyn A. Howell, William R. Jordan III, Michael J. Lemke, Jennifer Lyndall, Terry R. Robison, John Shuey, Carol Stronjny, and two anonymous reviewers for their review of earlier versions of selected chapters. We are grateful for the guidance and editing suggestions provided by Catherine Cocks and James McCoy of the University of Iowa Press and Faith Marcovecchio of Faith Marcovecchio Editorial, which enabled us to greatly improve the book. Thanks to Oxford University Press for permission to use an adapted version of Figure 6, originally published in Galatowitsch 2012 and Harwell et al. 1999b. We thank the Board of Directors of the Midwest–Great Lakes Chapter of the Society for Ecological Restoration for their encouragement and enthusiastic support of this project. On a more personal note, Chris thanks Michelle for her support during the years it took to get this book written and the larger Lenhart family for the experience of doing hands-on restoration work on our farmland in northwest Ohio. Peter thanks his wife Belynda for her love, support, and her motivational mandate: "Finish the book."

INTRODUCTION

CHRISTIAN LENHART AND PETER C. SMILEY JR.

The Midwest is unique among regions in the United States because of the combination of its industrial history, agricultural productivity, and natural features such as the Great Lakes. The decline of the automobile industry and other manufacturing enterprises that long made their home here left a legacy of pollution and abandoned infrastructure in a region now often called the Rust Belt. Assisting with the recovery of urbanized ecosystems damaged by industry requires creative thinking. So too does the severe ecosystem degradation caused by intensive row-crop agriculture in the Corn Belt in Ohio, Indiana, Illinois, and Iowa, as well as the southern parts of Michigan, Minnesota, and Wisconsin. Beyond the heavily used agro-industrial areas, much of the northern Midwest is still forested, presenting opportunities for larger scale natural area restoration.

We felt it was important to have a book focused solely on the Midwest because the practice of ecological restoration is often region-specific and in many cases site-specific. The unique combination of climate, soils, and biota in each region demands practices, solutions, and restoration approaches specific to that setting. The wealth of restoration history and current activity within the Midwest make a book focused on the unique attributes of ecological restoration in the Midwest–Great Lakes region, encompassing Ohio, Indiana, Michigan, Illinois, Wisconsin, Minnesota, and Iowa, an important addition to the literature.

Along with the regional focus, we are interested in highlighting the link between theory and practice in ecological restoration in the Midwest. The field, from its origins in the 1980s, presumed that the practice of restoration would contribute to the science and theory of restoration and vice versa (Jordan et al. 1987). There is a strong need to put theory into practice within ecological restoration, because ecological restoration is not simply an academic pursuit (Falk et al. 2006). Yet leading thinkers in the field have recognized that we lack guiding theories and principles. Likely the site-specific nature of restoration projects has impeded the development of universal principles. Without such principles, lessons learned from practice

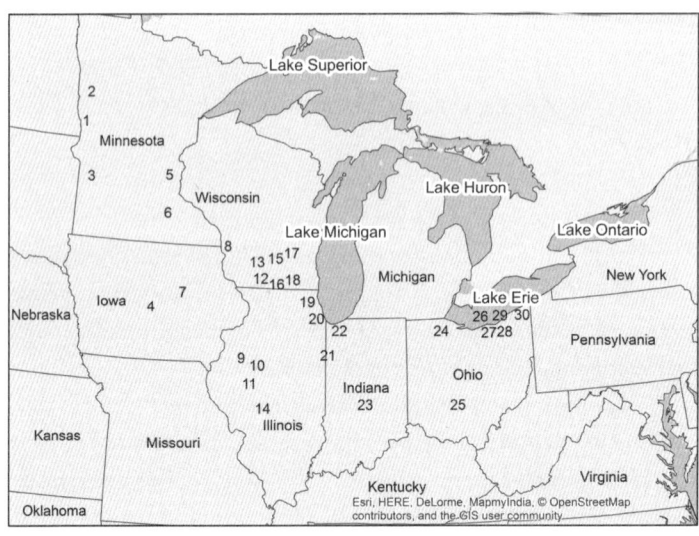

FIGURE 1. Locations of Midwest restoration efforts and research sites discussed in the book. Numerical codes for restoration efforts are 1: Lawndale Creek; 2: Wild Rice River; 3: Pomme de Terre River dam; 4: Story County Integrated Roadside Vegetation Management Program; 5: Cedar Creek Natural History Area; 6: Carleton College Arboretum; 7: Black Hawk County Integrated Roadside Vegetation Management Program; 8: Coon Creek Watershed; 9: Knox College; 10: Emiquon Complex; 11: Illinois River Biological Field Station; 12: Mount Vernon Creek; 13: Black Earth Creek; 14: Lincoln Memorial Garden; 15: University of Wisconsin–Madison Arboretum; 16: Yahara River; 17: Maunesha River; 18: Bark River; 19: Nippersink Creek; 20: Forest Preserves of Cook County; 21: Kankakee Sands Project; 22: Indiana Dunes; 23: Marian University Nina Mason Pullium EcoLab; 24: Maumee River; 25: Big Darby Creek; 26: Lake Erie; 27: Cuyahoga River; 28: Acacia golf course; 29: Holden Arboretum; 30: Ashtabula River.

are not necessarily relevant or applicable to similar projects done elsewhere, reducing progress in developing more effective restoration practices. Important lessons from practice should inform research and teaching as well. Unfortunately, the field of ecological restoration is too often divided, with practitioners on one side and scientists on the other. The existence of two terms (*restoration ecology* and *ecological restoration*) to define the different components of restoration illustrates the divide. Restoration ecology contributes science and theory in support of the practice of restoration, while ecological restoration refers to the design and implementation of projects (Falk et al. 2006). Throughout this book, we use *ecological restoration* to refer to both science and practice as a way to bridge the gap between these two components of this discipline and to promote the seamless link between theory and practice.

This book consists of ten chapters in three sections: the historical and theoretical underpinnings of ecological restoration, case studies that build on lessons from the past, and the future of ecological restoration in the Midwest. Additionally, thirty restoration efforts and research sites located throughout the Midwest are discussed within the book (Figure 1).

PART I HISTORICAL AND THEORETICAL UNDERPINNINGS

The Midwest has one of the longest histories of ecological restoration in the United States, dating back to the 1930s and earlier if one considers the precursory efforts to protect, repair, and restore damaged ecosystems in the late 1800s and early 1900s as part of the development of the field. The rise of ecological restoration in the Midwest corresponds with the rise of ecology there. Chapter 1 reviews the history of both, discussing the interrelationships between four pioneers in landscape architecture and ecology and how their joint and individual efforts at protecting natural areas and repairing damaged ecosystems contributed to the rise of modern-day ecological restoration. Additionally, Chapter 1 explores the evolution of ecological restoration in the Midwest by providing a detailed history of a current restoration project that began as a designed landscape implemented by famed landscape architect and conservationist Jens Jensen. This is an important contribution to the history of ecological restoration because our current understanding (Jordan and Lubick 2011) focuses only on ecocentric restoration efforts, that is, restoration efforts conducted only for the sake of the ecosystem and/or biota.

Theory and practice have evolved rapidly in recent decades by building on this strong historical legacy. Thus the field needs an up-to-date synthesis of theory and practice relevant to the issues that are distinctly important in the Midwest. The Midwest has played a key role in shaping both the scientific theory and the practice of ecological restoration in the United States.

In particular, its setting along the prairie/forest boundary in the nation's breadbasket has promoted the development of several key theoretical and practical aspects of the field. Chapter 2 reviews selected elements of ecological theory and practices that originated in the Midwest to highlight the contributions that the early pioneers in ecology and ecological restoration made to the field internationally. This review is particularly important because it identifies relevant theories and practices used in both terrestrial and aquatic ecosystems.

PART II CASE STUDIES – BUILDING ON LESSONS FROM THE PAST TO FORGE A NEW FUTURE

The site-specific nature of restoration makes it important to learn from case studies within a region. The case studies included in this book were selected to illustrate projects that integrate current theory into practice in the midwestern context. We selected case studies of prairie restoration (Chapter 3), floodplain wetland restoration (Chapter 4), stream restoration (Chapter 5), and urban ecosystem restoration (Chapter 6) to illustrate how some of the principles and issues discussed in Chapters 1 and 2 come into play in real-world applications.

Chapter 3 presents two prairie restoration case studies. One describes an attempt at restoring a high-quality historic prairie, and the second focuses on a more recent practical application of prairie restoration. The Curtis Prairie restoration at the University of Wisconsin–Madison Arboretum is one of the longest-term and best examples of the ideal in Midwest prairie restoration. It's considered ideal because from its beginnings, it stimulated research on prairie ecology and prairie restoration practices and highlighted the seamless nature of theory and practice in ecological restoration. The Curtis Prairie project is foundational and has stimulated new prairie restoration efforts in the region, including the Integrated Roadside Vegetation Management Program in Iowa, which is also presented in Chapter 3. This program takes a utilitarian approach and demonstrates how prairie restoration principles are being applied in roadside areas to opportunistically establish large areas of prairie vegetation in a landscape dominated by agricultural land use.

Wetlands have been the focus of restoration efforts since the 1950s, when people recognized that waterfowl were declining in the prairie pothole region of Iowa, Minnesota, and the Dakotas. In more recent years, wetlands have been restored increasingly for their ecological functions and values. Chapter 4 focuses on the Emiquon Floodplain wetland project along the Illinois River, which is one of the largest riverine floodplain restoration projects ever undertaken in the Midwest. It is a multipurpose restoration for water quality, flood retention, and wildlife habitat. Chapter 4 provides an overview of precursory floodplain wetland restoration efforts along the Illinois River and then highlights how these findings were used to design and implement the ambitious adaptive restoration approach now being used.

Chapter 5 presents case studies of stream restoration projects involving the use of natural channel design to assist with the recovery of Midwest streams damaged by channelization and dams. The restoration of Lawndale Creek in Minnesota represents a holistic effort in a formerly channelized headwater stream in the Red River basin. Dam removal has risen to the forefront of stream restoration work in the past decade because restoring connectivity is important for the recovery of many aquatic species. However, dam removal is more complex than simply reestablishing connectivity, as it also involves dealing with a host of issues related to sediment and vegetation management. The six dam removal projects discussed in Chapter 5 highlight this complexity.

The Rust Belt offers endless opportunities for cleaning up degraded urban ecosystems and finding ways to beneficially use space in less populated urban centers. Chapter 6 reviews the challenges and opportunities for such projects and discusses case studies of the restoration of the Ashtabula River and natural areas on the Acacia golf course in northeastern Ohio. Both case studies highlight that urban ecosystem restoration involves the restoration of multiple ecosystem types in a highly altered landscape, which compounds the complexity of these restoration efforts.

PART III THE FUTURE OF ECOLOGICAL RESTORATION IN THE MIDWEST

Ecological restoration has evolved from a focus on plant community structure in early prairie restoration projects to a focus on larger, landscape scales and the establishment of ecosystem functions and services. While invasive species have long been a major issue, new species introductions and range expansions continue to test our knowledge and ability to manage their negative impacts. Climate change has arisen as a major concern in the past decade and is a dominant research topic across a wide range of basic and applied fields of science. Although climate change has not had a major influence on the practice of ecological restoration in the past, it is increasingly playing a role in project planning and implementation. Finally, the functional, pragmatic concept of ecosystem services is becoming more and more important in ecological restoration, as evidenced by the focus at national conferences and in recent publications on sustainability and human/nature interactions (CBD 2011; McDonald et al. 2016). The use of practices that focus on restoring ecosystem services is especially prevalent in Midwest agricultural watersheds, where the main goal is water quality improvement.

Chapter 7 demonstrates that climate change and altered disturbance regimes will strongly influence the goals of restoration and management projects in a variety of ways. As the climate shifts toward warmer temperature regimes, biomes and ecosystems will shift northward, raising issues of species migration and the need for connectivity and corridors for movement of plants and animals (Galatowitsch et al. 2009). If the climate of northern locations changes too quickly, then less adaptable organisms will be reduced or potentially go extinct. Additionally, warmer air temperatures will lead to warmer water temperatures, which will stress upland plant communities and aquatic ecosystems, producing changes in biogeochemistry and nutrient cycling. Mitigation of climate change impacts will require increased carbon storage efforts, including prairie restoration and reforestation efforts. The example of current restoration efforts in Indiana forest and savanna ecosystems demonstrates how climate change adaptations can be incorporated into restoration design.

The management of invasive species is the focus of Chapter 8. The introduction of nonnative species around the globe is of growing importance, with impacts estimated at billions of dollars (Pimentel et al. 2005) and growing each year. Precautionary practices and controls are thought to be the most effective approaches in the short term. For example, Canada thistle (*Cirsium arvense*) is extremely problematic once it has taken over a wetland or prairie restoration project, but it may be controlled if kept in check from the start. Invasive aquatic animals (e.g., Asian carp [*Hypophthalmichthys* spp.], zebra mussels [*Dreissena polymorpha*], sea lampreys [*Petromyzon marinus*]) and invasive terrestrial insects (e.g., emerald ash borer [*Agrilus planipennis*]) are equally problematic because currently we lack effective ways of controlling them and they have widespread ecosystem impacts. Current research focuses on not only management and control but also screening of species that are not yet problematic in order to identify characteristics that make species invasive and destructive. Those that change the physical environment, thus undermining the viability of coexisting native species, are likely to have the greatest impact. Chapter 8 provides key strategies for developing a common vision for future invasive species prioritization and control in the Midwest.

Ecological restoration has traditionally focused on natural areas. As these decline in abundance, future restoration efforts within semicultural ecosystems, such as agricultural watersheds, will be critically important across large parts of the Midwest. Additionally, there is an increasing need to use ecological restoration to address the impacts of agriculture on Midwest watersheds, because traditional management efforts have not yielded the expected benefits. Chapter 9 shows how ecological restoration fits with other watershed management approaches and how incorporating it with existing approaches will assist in developing more effective watershed restoration strategies. This chapter also reviews five theoretical frameworks developed for designing watershed restoration strategies to identify the theoretical framework (Treatment Train Approach) most likely to lead to holistic strategies that address (1) altered hydrological regimes; (2) increased sediment, nutrient, and pesticide loadings; and (3) degraded physical habitat, which is prevalent within Midwest agricultural watersheds. The Treatment Train Approach highlights how traditional and emerging watershed management practices can be part of future restoration strategies that take into account the impacts and opportunities of all components of agricultural watersheds. The watershed-scale approach is critically important because typical stream restoration efforts focus only on the streams themselves and/or the riparian corridors.

Our concluding chapter identifies the links between key concepts and practices and discusses our vision for the future of ecological restoration in the region, which holds much promise even in light of the emerging threats to ecosystem integrity. Many areas related to the science and practice of ecological restoration still need improvement and further development. While the Midwest has a strong legacy of restoration experience to build on, there is also a legacy of ecological degradation that must be addressed in restoration. Ideally, ecological restoration seamlessly integrates theory and practice. Scientists and practitioners need to continue collaborating to meet this common goal, which will benefit the field as well as the degraded ecosystems in the region.

PART

ONE

Historical and

Theoretical

Underpinnings

CHAPTER ONE

DISCOVERING THE ROOTS OF ECOLOGY AND ECOLOGICAL RESTORATION IN THE MIDWEST

PETER C. SMILEY JR., DAVID P. BENSON,
AND JOHN A. HARRINGTON

Ecology is the field of science devoted to the study of the interrelationship of life and its biotic and abiotic environment. The field officially began in the United States with the organization of the Ecological Society of America in 1915 (Burgess 1976). However, individual scientists were conducting ecological studies decades earlier. Biologists, amateur naturalists, and explorers made observations on the occurrence and abundance of the flora and fauna in the 1800s (McIntosh 1986). Ecology originated in Europe, but its rise was rapid in the United States in the early 1900s, and American ecologists were soon recognized as leaders in this new subdiscipline of biology (McIntosh 1986). Since its inception, ecology has gone from mostly a descriptive natural history approach to the quantitative, hypothetical, deductive approach that dominates the field today (McIntosh 1986).

Ecological restoration is the science and practice devoted to assisting the recovery of an ecosystem that has been degraded, damaged, or destroyed (SER 2004). Similar to the field of ecology, a few individuals and organizations began practicing ecological restoration well before it formally organized as a professional discipline (Jordan and Lubick 2011). Although the practice garnered a lot of interest in the late 1800s and early 1900s, it went on hiatus for much of the mid–twentieth century before reemerging in the late 1900s (Jordan and Lubick 2011). The field of ecological restoration began to mature rapidly in the 1960s, which coincided with the rise of the environmental movement and the maturation of the field of ecology. In 1988, it became a formal discipline with the establishment of the Society for Ecological Restoration (Jordan and Lubick 2011). In light of increasing concerns over habitat loss, climate change, and decreasing biodiversity, the interest in ecological restoration continues to rise and its application is currently a multi–million dollar industry (BenDor et al. 2015).

The definition of *restoration* has been debated hotly since the 1980s, and the Society for Ecological Restoration has altered how it is defined multiple times (Higgs 2003). Early definitions centered on a return to historical conditions, but the current definition focuses on the repair and recovery of degraded ecosystems with the recognition that while historical conditions can serve as a guide, the final outcome might be different physically and biologically from what existed previously. Additionally, ecological restoration in its strictest sense is an ecocentric approach having an intent to benefit the biota and the ecosystem being restored (Jordan and Lubick 2011). The ecocentric approach separates ecological restoration from other management types (i.e., forestry, wildlife/fisheries management, environmental management, landscape architecture) that are guided primarily by anthropogenic concerns and interests. The most current understanding of the history of ecological restoration (Jordan and Lubick 2011) is limited because it focuses only on the history of the field based on ecocentric restoration efforts. Historical restoration efforts and other types of management projects having anthropogenic objectives represent the precursor to ecological restoration and can result in what can be considered restoration. Thus our understanding of the historical trends in ecological restoration will be expanded by considering the role that these historical precursory restoration efforts played.

Similarities between ecology and ecological restoration are apparent, as they involve working with living organisms that are influenced simultaneously by multiple abiotic and biotic factors operating at a wide range of spatial and temporal scales. Both are interdisciplinary fields that use information from various physical and biological sciences to achieve their goals. Ecology is a scientific discipline that focuses on increasing our understanding of the world, while ecological restoration is the application of ecology to the repair of degraded ecosystems. Restoration ecology, which is the science of ecological restoration, focuses on the use of the scientific method and ecology to increase the effectiveness of the practice of ecological restoration.

Ecological restoration is more than just restoration ecology and the practice of restoration. It encompasses all ideas, concepts, and practices from the life and physical sciences, economic fields, social fields, political fields, and art that are relevant to ecosystem restoration (Higgs 2003). While ecological restoration is influenced by many disciplines, the link between ecology and ecological restoration is clear. Furthermore, many key historical events in these fields occurred in the Midwest, but their historical links have not been explored in detail. Our objective is to expand our understanding of the history of ecological restoration by synthesizing the contributions of historical precursory restoration efforts resulting from the common rise of ecology and ecological restoration in the Midwest. We focus on the interrelationships and accomplishments of four pioneers in ecology and landscape architecture from the late 1800s to the 1950s to illustrate their contributions to ecological restoration and the intermingled history of ecology and ecological restoration in the region. We further highlight the importance of historical precursory restoration efforts by detailing the history of a current restoration project that began as a designed landscape implemented by famed landscape architect and conservationist Jens Jensen.

CULTURAL VIEWS

We begin with an overview of the cultural perspective of midwesterners in the early 1800s. In this period, Americans were settling the Midwest. As new settlers explored and moved west, they encountered prairies, forests, lakes, and streams in relatively pristine conditions. The hunter-gatherer lifestyle of Native Americans had likely altered the habitat, but these changes were small in scale and less intense than the changes that occurred with agrarian settlers and the coming of the Industrial Revolution. The early settlers focused on obtaining the resources that they needed for survival and establishing their homesteads without concern for the impact of their activities on the environment. Technology improved as the region became more populated, which made travel and obtaining needed resources easier (Kline 2011). Agriculture became increasingly mechanized with increasing industrialization, resulting in increasing amounts of land being put into production (Kline 2011). The belief that nature existed to benefit humanity encouraged the exploitation of natural resources during this period (Kline

2011). Additionally, increasing population size and the closing of the western frontier led to increasing urbanization and, in combination with growing industrialization in Midwest cities, crowding, air and water pollution, and destruction of natural landscapes (Kline 2011). By the late 1800s individuals and groups were becoming concerned with the habitat destruction and degradation they were observing (Egan and Tishler 1999; Grese 2011). Many of these conservation-minded and progressive individuals felt nature could provide a respite for the ills of modern life (Jensen 1939; Engel 1983; Egan and Tishler 1999; Grese 2011). This feeling of ongoing social and environmental decline provided the initiative to beautify, preserve, and restore (Miller 1915). This impulse ultimately led to the rise of ecology and ecological restoration in the Midwest (Egan and Tishler 1999).

IN THE BEGINNING

In the late 1800s and early 1900s science and practice were not separated into the fields of biology and ecology as they currently are. The sciences generally were less specialized as many early naturalists, biologists, and ecologists worked with multiple taxonomic groups and different ecosystem types over their careers (Croker 1991, 2001). In contrast, today many scientists specialize in a single ecosystem and/or taxonomic type. As a result, the early ecological restoration attempts in the Midwest were undertaken by biologists, botanists, ecologists, and landscape architects with broad interests and experiences (Jordan and Lubick 2011). For example, mammalogist Benjamin Patterson Bole Jr. and ecologist Arthur B. Williams conceived and carried out the restoration project at Holden Arboretum in Ohio in 1930 (Jordan and Lubick 2011). Biology professor Harvey Stork and grounds superintendent D. Blake Stewart planted native trees and shrubs on the grounds of the Carleton College Arboretum in Minnesota beginning in 1926 (Jordan and Lubick 2011). Wildlife ecologist Aldo Leopold, horticulturalist and landscape architect G. William Longenecker, and botanist Norman Fassett supervised the initial restoration projects at the University of Wisconsin–Madison Arboretum in the 1930s (Sachse 1974). Foresters, wildlife managers, and fisheries managers working for state and federal agencies carried out other early precursory restoration efforts. Early landscape architects were particularly influential through their on-the-ground activities and writings. Landscape architects associated with the prairie style of landscape design promoted the use of a naturalistic style of landscaping reflective of midwestern and regional plant species and materials. Many of these individuals learned to identify native plants and understood their natural history in order to use them in their landscape designs (Grese 2011). Likely the common link among the individuals who contributed to the rise of ecology and ecological restoration in these early days was a shared love of the flora, fauna, and the landscape of the Midwest.

RELATIONSHIPS BETWEEN PIONEERS IN LANDSCAPE ARCHITECTURE AND ECOLOGY

Ecological restoration is the application of ecology and as such must go beyond the science and encompass individuals with expertise in terrestrial and aquatic ecology as well as in the design and implementation of landscape alteration projects. Jens Jensen (1860–1951), Henry Chandler Cowles (1869–1939), Stephen Forbes (1844–1930), and Victor Shelford (1877–1968) were all pioneers in ecology, ecological restoration, and conservation. Jensen was by profession a landscape architect, and Cowles, Forbes, and Shelford were ecologists. The relationships between these four individuals during their careers highlight the contributions of precursory restoration efforts and the rise of ecology and ecological restoration within the Midwest. The following synthesis draws on Croker (1991), Grese (1995), Engel (1983), Croker (2001), Cassidy (2007), and Grese (2011).

Cowles was a faculty member at the University of Chicago in the early 1900s. Cowles conducted his dissertation on the successional dynamics of the vegetation at the Indiana Dunes. This work was widely considered a significant contribution to ecology through its contribution to the theory of succession, and it was considered instrumental in establishing ecology within the United States. Following the completion of his PhD, Cowles was hired as a laboratory assistant to develop the ecology curriculum at the University of Chicago and he worked his way up to the rank of full professor. Cowles was a gifted and dedicated teacher who donated his time generously to a number of conservation-oriented organizations. It was through his volunteer activities with the Geographic Society of Chicago that he met Jensen in 1901.

Cowles and Jensen became friends and took trips into the Indiana Dunes and other natural areas for the collection and identification of plants. In 1905 Jensen organized the Playground Association of Chicago, and he and Cowles served on the committee that sponsored the association's Saturday afternoon walks. These trips to the natural areas in and around Chicago were intended to provide the poor with an opportunity to experience natural areas and increase their awareness of these sites. The walks became so popular, especially among the middle class, that Jensen, Cowles, and Chicago reformer Thomas W. Allinson started a new organization called the Prairie Club to administer them. The Prairie Club was primarily a social organization for professionals, but for many years it functioned as the midwestern counterpart to the Sierra Club in the western United States. In 1913, Jensen, Cowles, and Allinson, along with sixteen other prominent citizens, organized Friends of Our Native Landscape, which was a society formed to promote life in the open and to preserve examples of native landscapes that were fast disappearing as a result of industrialization. Jensen was the president, Cowles served as vice president, and Forbes was a member of this group. Friends of Our Native Landscape was foremost among the five other Chicago-area conservation groups that were organized by Jensen, Cowles, and Allinson together or separately.

One of the most significant accomplishments of these conservation organizations was their contribution to the establishment of Indiana Dunes State Park and Indiana Dunes National Lakeshore. The Indiana Dunes, located between Calumet and Michigan City, Indiana, is a unique ecosystem in the Midwest that was threatened by the increasing industrialization and urbanization that was occurring in the early 1900s. The importance of the Indiana Dunes for ecologists internationally is highlighted by the insistence of the members of the 1913 International Phytogeographic Excursion of the United States that they visit this ecosystem. Cowles and Shelford led a tour of the dunes for this group of international scientists, which included famed ecologists Arthur Tansley and Frederic Clements.

The effort to protect the Indiana Dunes began in the early 1900s, when Jensen began traveling through the Chicago area promoting the dunes' importance and beauty. Later, the Prairie Club, Friends of Our Native Landscape, and the National Dunes Park Association, all formed by Jensen, Cowles, and Allinson, assumed the primary role in advocating for the Indiana Dunes. These early conservation efforts to protect the dunes were unique because they involved prominent individuals from the arts, sciences, and public service. Despite the ecological significance and the public popularity of the Indiana Dunes, it took a decade of effort by numerous individuals and conservation organizations to establish Indiana Dunes State Park in 1925 and then another forty-one years to establish Indiana Dunes National Lakeshore to ensure the protection of this unique ecosystem from urbanization and industrialization.

Friends of Our Native Landscape published a publicity piece titled *Proposed Park Areas in the State of Illinois* in 1921. This publication described the scenic beauty and historical interest of natural areas within Illinois that needed to be preserved. These natural areas were identified by Cowles, Jensen, and Forbes using information compiled through the Illinois State Forestry Survey and fieldwork conducted by Cowles. This significant publication led to the protection and management of many of these natural areas, as they became state parks or forest preserves.

Cowles and Jensen collaborated for at least fifteen years. Cowles contributed to Jensen's understanding of plant ecology, the role of individual species within the larger plant community, and succession within plant communities. Jensen's increased knowledge of plant ecology helped him implement his landscape designs within Chicago-area parks and private residences. Jensen was a noted leader of the prairie style of landscape design and promoted the use of native plants. He used relatively pristine native landscapes as inspiration for his designs, a precursor to today's restoration practice of designing restoration projects based on the physical and biological conditions found at nearby unimpacted reference sites (for more on this reference ecosystem concept, see Chapter 2). Jensen never intended to duplicate nature, but instead he tried to capture the spirit of the native landscapes in his projects. Jensen's design of the Lincoln Memorial Garden in the mid-1930s is reflective of modern-day ecological restoration. Most of the seedlings and transplants predominant in his design were planted by volunteers of all ages, presenting the idea that restoration can be an ecological, social, and educational activity. Jensen intended for succession to occur and for the vegetation to change over time. Even Jensen's work on private estates reflected succession and transformation of existing landscapes to one designed with native landscapes in mind. Specifically, the intent of his designs was to foster a love for the native landscape and ultimately assist with conservation efforts.

In 1907, Cowles became a charter member of the Illinois Academy of Science, and he eventually served on its Committee on Ecological Survey. Stephen Forbes was the chair of this committee, formed to improve the understanding of Illinois forests and to develop policy for public and private forested lands. Cowles was an advocate for responsible forest policy, and as early as 1911 he recommended that "bits of forests here and there gradually be restored to the primitive wilderness and beauty of the forests of pioneer days" (Cassidy 2007). Cowles's involvement with this committee is notable because it represents one of the earliest documented interactions between these two early ecologists. In the summer of 1918, Cowles conducted surveys in northern Illinois documenting the extent and types of forested habitats in the state. The results of these surveys, as well as those conducted by Forbes's Illinois Natural History Survey colleagues, were summarized in two reports published in 1921 and 1923. The 1923 report was used as the impetus to establish a forestry division within the University Experiment Station. The 1923 report also contributed to a 1926 act that authorized the Illinois Department of Conservation to establish state forests as fish and wildlife sanctuaries and public parks, and to develop tree nurseries to support reforestation efforts.

Forbes was a versatile ecologist who worked with birds, terrestrial insects, plankton, aquatic macroinvertebrates, and fish. He conducted pioneering research in economic entomology, economic ornithology, limnology, and ecology, and was an exceptionally productive scientist, writing over 200 publications during his career. As a result of his foresight and efforts, he established the Illinois Natural History Survey, which is one of the premier state natural history surveys in the nation. In 1894, Forbes requested and received funds from the Illinois legislature to establish the Illinois River Biological Field Station in Havana, Illinois. As part of this proposal, he indicated that he would compare the chemical and biological conditions of the river before and after the opening of the Chicago Drainage Canal into the Illinois River. The drainage canal was expected to reverse the flow of the Chicago River and water from Lake Michigan, and transport diluted sewage, stockyard, and industrial wastes from Chicago into the Des Plaines River and then the Illinois River. Forbes's research through the Illinois River Biological Field Station documented the physical, chemical, and biological impacts of the canal. This scientific evidence resulted in required construction of two sewage plants to treat wastewater coming from Chicago prior to its entry into the Illinois River. This represents one of the first applications of ecology research results to assist with the repair of a degraded river in the United States. Additionally, as part of this research effort Forbes and Robert Richardson developed a water pollution index that used benthic aquatic macroinvertebrates as indicators of the water quality of the river. This was the first water pollution index in the United States and a precursor to widely used quality indexes developed by Illinois scientists, such as the floristic quality index developed by Morton Arboretum scientists in the 1970s (Swink and Wilhem 1979) and the Index of Biotic Integrity developed by a University of Illinois scientist in the 1980s (Karr 1981, see Chapter 2). These types of indices help quantify the quality of remnant ecosystems and serve as criteria for monitoring the impacts of a restoration project.

Victor Shelford was a distinguished ecologist whose work helped set the foundation for modern ecology. Over the course of his career, Shelford conducted field and laboratory research involving community ecology, ecotoxicology, and population ecology in freshwater and marine ecosystems. He obtained his PhD from the University of Chicago, and his dissertation research focused on tiger beetle (subfamily Cincindelinae) life history and larval habitats. Cowles was one of Shelford's graduate school mentors and is credited as the teacher who influenced him the most. Cowles's influence likely inspired Shelford, as a graduate student, to conduct a study of the association of tiger beetles with plants on the Indiana Dunes. After completing his dissertation, he became a professor at the University of Chicago and then later was recruited by Forbes to become a professor at the University of Illinois and a biologist with the Illinois Natural History Survey.

In December 1914, a committee was formed to organize an American ecological society to provide an opportunity for joint gatherings of plant and animal ecologists and to organize summer field trips for its members (Shelford 1917). Shelford was elected president of the organizing committee and Cowles was its treasurer-secretary (Shelford 1917). In December 1915 in Columbus, Ohio, fifty members approved the proposed constitution and the Ecological Society of America (ESA) was formed (Shelford 1917). Shelford was elected ESA's first president. Later, Cowles would serve as president (in 1918), and Forbes was the president in 1921.

In 1917, Shelford proposed the formation of an ESA Committee for the Preservation of Natural Areas and was appointed chair of the committee. Cowles and Forbes were also members of this committee. The committee published a milestone report in 1933 that identified natural areas in the United

States that needed to be preserved and their current condition. This report also provided management recommendations that exhibited exceptional foresight (Grumbine 1994). Remarkably, the report listed modern management recommendations for natural areas such as (1) the importance of protecting individual species and ecosystems; (2) the expansion of natural reserves to match the species' needs; (3) ensuring that preservation efforts protect a representation of ecosystem types; (4) accounting for natural disturbances; (5) and the use of buffers to protect core habitats within natural areas (Grumbine 1994).

Shelford from the beginning felt strongly that ESA needed to be actively involved with conservation of natural habitats, so he contributed time and personal funds to the work of the Committee for the Preservation of Natural Areas. In contrast, many in the leadership of ESA did not think that it was appropriate for a learned scientific society to be actively involved with conservation. This difference in philosophy eventually led to the termination of the preservation committee. In response, Shelford organized the Ecologists' Union in 1946 to continue the committee's work separate from ESA. In four years the Ecologists' Union had grown from 83 members consisting only of ESA members to 294 members consisting of both scientists and nonscientists. In 1950, the Ecologists' Union reorganized and changed its name to The Nature Conservancy and Shelford served on its board of governors. The Nature Conservancy has since become a leader in preserving and restoring terrestrial and aquatic ecosystems throughout the world.

In conclusion, the relationships between Cowles, Jensen, Forbes, and Shelford highlight the beginnings of ecology and ecological restoration in the Midwest. Their individual and joint accomplishments resulted in the protection of important natural areas, the development of a quality index, and the organization of state, national, and international organizations that are currently contributing to modern-day ecology and ecological restoration.

THE HISTORY OF MARIAN UNIVERSITY'S NINA MASON PULLIAM ECOLAB

The Initial Design

The history of Marian University's Nina Mason Pulliam EcoLab illustrates how a precursory restoration effort (i.e., designed landscape) laid the foundation for a modern-day ecological restoration project. In 1910, James Allison, a prominent Indianapolis businessman, purchased farmland in Indianapolis, Indiana, that would become his Riverdale estate (Figure 2). At the time of purchase, the northern portion of this 0.26-kilometer-square property consisted of row crops, with pasture to the south. The property contained a portion of Crooked Creek, located about 1.6 kilometers upstream from the White River. The meandering of Crooked Creek and the postglacial flooding of the White River watershed created an eighteen-meter-high bluff on the southern edge of the property. Construction of Allison's home at the top of this bluff began in 1911 and was completed in three years. After construction, the home, with its modern conveniences, became known as the "House of Wonders."

The same postglacial flooding that carved the bluff also exposed a soil layer impervious to infiltration. Water percolating through the soil hit the impervious layer and then eventually resurfaced within the Riverdale estate and the surrounding area. Just north of the mansion, at the base of the bluff, were several springs and wetlands that were drained by the farmer with subsurface tile drains. The farmer also used the springs to his advantage by building a groundwater-fed livestock watering trough.

Allison hired Jens Jensen in 1910 to design the landscape for his Riverdale estate. Jensen made several scouting trips to Riverdale as he began developing a plan. During these trips, he mapped the site and took photographs to document the existing conditions. Jensen's topographical maps documented the locations of the subsurface tile drains installed by the farmer, and his later design drawings indicated the tiles were used to maintain consistent pond levels. Jensen manipulated the hydrology across a significant portion of the estate. He built three dams within Crooked Creek that were lined with river rock. They contained sluice gates to manipulate the amount of water flowing over the dam to maintain an aesthetically pleasing flow. Jensen also armored a portion of Crooked Creek with concrete sides and created a prairie pool that was 0.3 to 1.2 meters in depth to evoke his idealized image of a wide, shallow prairie river. The retaining wall used moraine rock that probably originated from Crooked Creek itself, rather than the stratified limestone he used in his other projects (Dodson 1998). Jensen also created a series of five shallow spring-fed ponds along the bluff to the south, just below Allison's mansion (Figure 2). Allison's head groundskeeper indicated these ponds were "made the hard way—with teams and slip

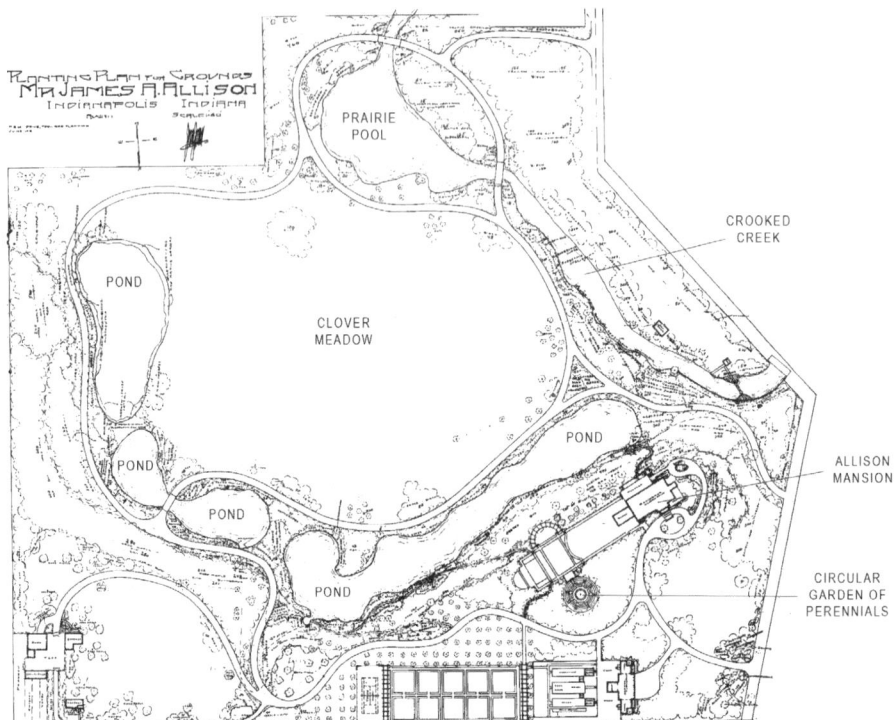

PRAIRIE
POOL

CROOKED
CREEK

POND

CLOVER
MEADOW

POND

ALLISON
MANSION

POND

POND

POND

CIRCULAR
GARDEN OF
PERENNIALS

FIGURE 2. Jens Jensen's planting plan for the grounds of the Allison estate in Indianapolis, Indiana. Only key features within the planting plan are highlighted. This site became part of the Nina Mason Pulliam EcoLab at Marian University, a laboratory for ecological restoration.

scoops" (Heidkamp 1952). The ponds were located in areas that would likely have been extremely wet with spring water even in the presence of the subsurface tile drains. Three of the ponds were dredged to a depth of 0.9 to 1.2 meters, while the other two were shallower wetlands located on the western part of the property.

Jensen designed a more formal landscape with distinct groupings and geometric patterns near the mansion. Just west of the mansion, he installed a labyrinthine circular garden of perennials surrounding a fountain (Figure 2). Near this garden was his iconic council ring, which consisted of a seating area around a fire pit within a half-circle stone colonnade and pergola. Jensen built an outdoor amphitheater between the circular garden and the council ring. The stage was raised and flanked by limestone pedestals with a backdrop of eastern red cedars (*Juniperus virginiana*) and eastern white pine (*Pinus strobus*). A large grove of sugar maples (*Acer saccharum*) was planted west and south of the mansion.

The majority of the plantings near the mansion consisted of native species, although Jensen's source for nursery stock extended beyond the local area. The cedars and pines were brought in from southern Indiana and Wisconsin (Heidkamp 1952). White birch (*Betula papyrifera*) were planted between the maples and the cedars, and were one of the few plantings not native to central Indiana. However, white birch is native to northern Indiana and Illinois, and it occurs in the Chicago area. Jensen also planted privet (*Ligustrum* spp.) in the formal portion of the landscape.

Jensen designed a less formal, more naturalistic portion of the landscape north of the bluff. Here Jensen's skills as a prescient restoration practitioner are most evident. For example, Jensen's plan indicates that 600 sugar maples and 600 red oaks (*Quercus rubra*) were to be planted in a 152-by-23-meter stretch located high on the bluff. Then just downslope, 560 black walnuts (*Juglans nigra*) and 560 butternuts (*Juglans cinerea*) were to be planted in a 107-by-15-meter location. The plantings of enormous numbers of individual plants rather than just a few large specimens is indicative of Jensen's willingness to allow ecological processes such as selection, competition, and succession take place. Jensen also specified the planting of 20 downy hawthorns (*Crataegus mollis*), 100 flowering dogwoods (*Cornus florida*), and 36 pepperidges (*Nyssa sylvatica*) below the walnuts along the shore of one of the shallow ponds. While flowering dogwood is typically found in drier environments, downy hawthorn and pepperidge are common in wet woods and along aquatic habitats.

Jensen did not specify the planting of many forbs or any understory vegetation, except in his plantings around the ponds and in the riparian habitat adjacent to Crooked Creek. Jensen had several common native wetland forbs planted along the wetland edges, such as swamp rose mallow (*Hibiscus moeschettus*), spikerush (*Eleocharis* spp.), northern blue flag (*Iris versicolor*), sweet flag (*Acorus americanus*), and buttercup (*Ranunculus* spp.). Jensen's designs also called for planting narrowleaf cattail (*Typha angustifolia*) along the wetland edge. The planting of narrowleaf cattail as part of a restoration project would be a questionable choice today, but at the time it was considered a relatively uncommon native cattail.

In the sunny riparian habitat of Crooked Creek, Jensen used overlapping flights of wet prairie–type wildflowers such as marsh blazingstar (*Liatris spicata*), New England aster (*Symphyotrichum novae-angliae*), Canada goldenrod (*Solidago canadensis*), sweet black-eyed Susan (*Rudbeckia subtomentosa*), black-eyed Susan (*Rudbeckia hirta*), smooth beggarticks (*Bidens laevis*), purple coneflower (*Echinacea purpurea*), showy black-eyed Susan (*Rudbeckia speciosa* var. sullivantii), summer phlox (*Phlox paniculata*), and goldenrod (*Solidago* spp.). He also specified a few species typically found in drier areas, like smooth blue aster (*Symphyotrichum laeve*), Missouri evening primrose (*Oenothera macrocarpa*), and narrowleaf evening primrose (*Oenothera fruticosa*). New York American aster (*Symphyotrichum novi-belgii*) was also used, although its native range lies east of Indiana. Jensen's designs also specify planting *Lythrum grandiflorum* in the riparian habitat of Crooked Creek. This species name is no longer in use, and it is uncertain which species it represents. Unfortunately, it is suspected that this species is purple loosestrife (*Lythrum salicaria*), which is a nonnative plant brought to the Midwest at the turn of the twentieth century by the horticultural trade. However, *Lythrum grandiflorum* may also have referred to another native *Lythrum* species, such as winged loosestrife (*Lythrum alatum*) (R. Grese, personal communication).

Jensen's plantings within the Clover Meadow in the center of Riverdale (Figure 2) most likely consisted of nonnative turf grasses (family Poaceae) and clover (*Trifolium* spp.). Jensen at this point in his career felt that the illusion of prairie was more important than creating the actual thing. By 1916, Jensen would use prairie grasses and wildflowers in his open spaces and designate them "the prairie" in his landscape drawings (Dodson 1998). Jensen's use of nonnatives in the Clover Meadow may have also been influenced by Allison's plan for the area to be grazed by cattle.

Jensen finished the landscape design in 1912, and the landscape installation was completed shortly after. James Allison and his wife, Sara, moved into their summer estate at Riverdale in 1913, and they used the estate regularly from then until his death in 1928.

Disinterest and Neglect

After Allison's death, Riverdale was in limbo as Allison's mother, his first wife, and his second wife fought over the inheritance. Allison's mother eventually won the dispute and maintained control of the estate until she died in 1930. Otis J. Clemens began working for Allison as the head groundskeeper and manager of his twenty-two maintenance men in 1921 and lived in a residence at the west edge of the estate (Heidkamp 1952). It is assumed that Clemens continued to maintain Jensen's original plantings through this period.

The Sisters of Saint Francis at Oldenburg purchased Riverdale in 1936. The Sisters then moved Marian College from Oldenberg to Riverdale in 1937 as part of their efforts to provide a college education for women. Marian College was the first Catholic college for women in Indianapolis. In its first year at Riverdale, the entire college, which consisted of twenty-four students and sixteen instructors, was housed within the estate. When the Sisters purchased the property, they retained head groundskeeper Clemens in their service until 1952 (Heidkamp 1952). Due to cost, it is unlikely the Sisters were able to keep as many groundspeople as the Allisons had, and the more naturalistic portion of the property was likely unmanaged.

Ice-skating and rowboating in the ponds and wetlands created by Jensen were major forms of entertainment for students from the 1940s on. The ponds and wetlands were excellent venues for these activities, but they were small. These shallow aquatic habitats were perceived by the students as being clogged and unclean, which reflected the common public perception of wetlands at the time.

In 1946 the Sisters agreed to a plan to dredge the three main ponds to create one larger and deeper lake and then to use the dredge spoil to infill the shallowest two wetland ponds and create a playground area (Bohlen 1946). Ironically, the dredging created drainage problems and may have created more wetland area on the property than the original wetlands that were filled. A substantial portion of Jensen's Clover Meadow became wet and reverted to hardwood swamp over the next several decades. After the dredging and filling of the original wetlands was completed, the December issue of the student newspaper promoted the use of the new lake by encouraging students to include ice skates on their Christmas lists (Anonymous 1946). The new, larger lake was such a success that lights were installed around the lake within a decade to allow for ice-skating at night (Anonymous 1959).

The upper part of campus property underwent major changes as the college expanded. Clare Hall was built in 1949, and Marian Hall was constructed in 1953. To make room for Marian Hall and parking, a large portion of the estate was dismantled, including a long colonnade, greenhouses, and a portion of the formal gardens. Some of the demolition debris, including the columns of the colonnade, was dumped onto the flat spot located on the western portion of estate where the shallow Jensen wetlands were located. This area became known to the campus community as the Roman Ruins (Yarber 1959). Additionally, when the basements for Marian and Clare Halls were being excavated, their fill was dumped into this same area, further infilling Jensen's wetlands and increasing the ground level three to five meters above its original elevation.

The wetlands and the other habitats located below the bluff were likely used by biology professors as a learning environment for their students as early as 1937. Botany and Taxonomy of the Higher Plants were taught in the early 1940s, followed by Local Flora in the late 1940s and into the 1950s. It is impossible to imagine these classes would not have taken advantage of the variety of vegetation in Riverdale. In the 1960s, Ecology was added to the curriculum and was a required class for biology majors from the 1970s to 2006. The Sisters of Saint Francis undoubtedly used the natural habitats as course materials for their students. The Sisters also used them for their own Franciscan Summer Study, where Sisters engaged in graduate and undergraduate study that included aquatic biology.

In the mid-1950s, the sophomore class created a Campus Conservation Corps to beautify the campus. As part of these beautification activities, the Campus Conservation Corps uncovered old trails around the lake and up to the stone arch bridge. In 1955, 1,650 multiflora rose (*Rosa multiflora*) seedlings were requested through the Indiana Department of Conservation's Living Fence Program (Olivia 1955). Multiflora rose has since become a major invasive species throughout the Midwest. The Campus Conservation Corps also raked leaves and installed plantings, but it is unclear whether they planted within the wetlands and the other habitats located below the bluff (Anonymous 1956).

Renewed Interest

In 1972 Marian College was the recipient of a major grant from the U.S. Department of Health, Education, and Welfare to develop the wetlands of the Riverdale estate into the Marian College Wetlands Ecological Laboratory, which would serve as an outdoor learning environment for Marian College students and local high school and grade school students (Anonymous 1972). The Minnesota Environmental Sciences Foundation was hired to design the master plan, and they envisioned that the Wetlands Ecological Laboratory would provide unique opportunities for basic research and teacher training in helping people understand their relationship to the surrounding environment (Vogt and Thompson 1973). The master plan included detailed descriptions of twenty-six sites within the Riverdale estate that could be used for teaching topics such as forest understory succession, wildlife observation, long-term ecological studies, and stream studies. The master plan also included a trail designed to depict hardwood forest succession, recommendations for rejuvenating selected springs to enable them to be used for raising fish for restocking and study, clearing Crooked Creek of logs and litter, planting shade-tolerant native understory plants along the bluff for erosion control, and rototilling old trails to loosen soil followed by planting native understory plants. While the site plan was being prepared, students, the Biology Club, and other volunteers removed trash and debris from the site and worked at maintaining water flow through the ponds to the creek.

The Marian College Wetlands Ecological Laboratory was a priority for the college for several years after the 1973 proposal was completed. The laboratory had a director and was used to involve students in learning about their environment and in the development and upkeep of the ecological laboratory itself (Anonymous 1978). The college diverted drainage from parking lots and roadways into culverts and storm sewers to inhibit further erosion. Campus Operations also used telephone poles to block access to the area to prevent dumping. The laboratory produced two books of educational activities to help with environmental education efforts. It is unknown whether these books were ever used. Funds were never secured to complete the plans for the trail system and study sites, and the property underwent a second period of disinterest until the same ideas were resurrected in 2000.

Early in 2000, Marian College students started quietly protesting the dumping that was occurring in the wetlands by Campus Operations and neighbors. The students' interest started a movement to develop the area into a more useable space. Several volunteer days were planned, including one featuring the Marian College president and cabinet. A Conservation Biology class wrote small grant proposals to fund a service-learning project for 120 third- through seventh-grade students who helped remove debris and plant native plants. These activities attracted public interest, and Indianapolis mayor Bart Peterson visited the campus to see the progress of these efforts.

The local Audubon chapter provided initial funding for habitat improvement, Asian bush honeysuckle (*Lonicera* spp.) removal, and planting a small prairie patch in the former Clover Meadow. This initial effort was followed by major funding from several sources, most notably the Nina Mason Pulliam Charitable Trust. Initial plans were drawn up, loosely following the recommendations from the 1973 Minnesota Environmental Sciences Foundation report (Vogt and Thompson 1973). Smock Fansler Corporation was hired to build the infrastructure of what was then being called the Marian College EcoLab.

The Jensen design for the landscape of Riverdale was rediscovered in a drawer in Marian College Campus Operations. Landscape architect Kevin Parsons, Tom Fansler III, Spencer Goehl from the restoration firm Eco Logic LLC, and biology professor David Benson analyzed the historic plans and found the road system was similar to the planned EcoLab trails. At the time, it was unclear whether the Jensen design had ever been implemented. While scouting out the location for a trail, the large concrete bridge at the west end of the pond was discovered under 0.6 meters of silt. Later, stepping-stones were found in the water along the south edge of the pond. With these finds it became obvious that the Jensen landscape had been built and the direction of the EcoLab project needed to change to include ecological restoration and the restoration of significant historical features.

The approach for historical and ecological restoration of the vegetation in the Jensen-designed area was simple. The Jensen design was followed if the plants specified for a particular area still appeared appropriate when taking the current environment, climate, and knowledge of ecology into account. For example, along the 213 meters of shoreline on the north edge of the main pond, Jensen originally specified the planting of 75 pepperidges, 120 smooth roses (*Rosa blanda*), 140 dogwoods (*Cornus* spp.), 335 crimsoneyed rosemallows (*Hibiscus moscheutos*), 36 hawthorns (*Crataegus* spp.), 42 white birch, and 12 ash (*Fraxinus* spp.). For the dogwood, ash, and hawthorn plantings, pagoda dogwood (*Cornus alternifolia*), green ash (*Fraxinus pennsylvanica*), and downy hawthorn were chosen because all three were commonly used by Jensen. With the exception of white birch, all of these plants are natives that do well in medium to wet soils and full sun. Although it changed the aesthetics slightly, river birch was used instead of white birch because white birch is not native to central Indiana and would be unlikely to survive the heat. Also, the number of individuals planted in the EcoLab was about ten times fewer than the number specified by Jensen. This may have altered the aesthetics, but the new plantings consisted mostly of 1.5-meter-tall plants, likely much larger than the individuals Jensen planted. The shoreline of the north edge of the main pond was then seeded and planted with a mixture of wetland plant species. Jensen did not always specify plants as understory, but in this case he gave a general list of species to be used along the shorelines of the ponds. The species planted contained all species specified by Jensen and several others to increase the diversity of the community.

Asian bush honeysuckle was noted within the EcoLab area as early as the 1940s, although it is unclear whether these early observations referred to Japanese honeysuckle (*Lonicera japonica*) or Amur honeysuckle (*Lonicera maackii*). By 2000, Amur honeysuckle was the dominant shrub within the EcoLab and was so abundant that it forced those marking the locations of the EcoLab trails to crawl through the shrubs. Other invasive shrubs and vines such as privet, multiflora rose, Japanese honeysuckle, common buckthorn (*Rhamnus cathartica*), Oriental bittersweet (*Celastrus orbiculatus*), and Norway maple (*Acer platanoides*) occupied the areas where Amur honeysuckle coverage was less than complete. Likely the majority of the EcoLab was overrun by invasive shrubs and vines by the turn of the twenty-first century. In 2001, the first major restoration action for the EcoLab was to remove the invasive shrubs and vines. Immediately afterward a mixture of silky wildrye (*Elymus villosus*), Canada wildrye (*Elymus canadensis*), and Virginia wildrye (*Elymus virginicus*) was planted in the newly opened areas, and then tree and shrub seedlings following Jensen's plan as described above were planted.

Restoration of the EcoLab also involved the removal of large quantities of fill to restore the wetlands in their original locations as specified by Jensen. The main pond was recontoured using this fill to more closely match the initial Jensen-designed shape of ponds one, two, and three. The recontouring of the main pond also allowed the trails to be installed in the original conformation of the roads and footpaths at Riverdale.

The Restoration Legacy of Jensen's Landscape Design

Prior to implementation of Jensen's landscape design, the Riverdale estate consisted of agricultural land devoid of almost all woody plants. There is no doubt that Jensen was attempting to create a particular aesthetic in his design. Notably, he chose plant species quite similar to those that many restoration practitioners would choose today, and his design called for groupings of native plants as they would be found in nature, fitting the region, microclimate, and soils of the area being planted (Grese 1992).

Of the many thousands of trees that were specified in Jensen's landscape design of 1912, only a small number are present within the EcoLab. A 2002 study documented the presence of only 116 trees of the correct species and approximate location specified in Jensen's original design. Of these trees, only twenty-three oaks (*Quercus* spp.), twenty-three sugar and black maples (*Acer saccharum, Acer nigrum*), fourteen eastern red cedars (*Juniperus virginiana*), and twelve eastern white pines (*Pinus strobus*) were approximately the right age to have been planted by Jensen. Approximately half (thirty-six) of the Jensen-aged trees that remained occurred within the naturalistic portion of Riverdale and therefore were not likely to have been subjected to pruning or trimming by the ground crews. Additionally, twenty-three shrubs were found of the correct species in the correct approximate location.

Much of the loss of Jensen's original plantings may be due to human impacts after the purchase of the property by Marian College in 1937, such as the dumping of fill into one of Jensen's wetlands, the recontouring of the main pond, and extra fill that was pushed over the bluff. These actions would have smothered sensitive root systems and caused stems to rot. Other plants were likely lost to natural ecological processes such as disease, climatic conditions, and succession. For example, it is likely white birch expired due to disease and climatic conditions because it often succumbs to bronze birch borer (*Agrilus anxius*) infestation and the heat. Additionally, the staghorn sumac (*Rhus typhina*) planted alongside sugar maples were probably shaded out by their larger neighbors. Other smaller species of shrubs and trees might have been lost as a result of the invasion of non-native shrubs and vines.

Assuming the landscape was installed as designed, which was not always the case (Grese 1992), only a small percentage of Jensen's original plantings survived (Benson 2004). However, the resulting native plant community of the EcoLab was in remarkably good shape in 2001 and exhibited a 3.3 mean coefficient of conservatism and a high floristic quality index score (53.5) (based on a recalculation of data obtained by Tungesvick [2003] using Indiana coefficients of conservatism). The EcoLab's floristic quality index score is a good indication of the success of Jensen's design, especially considering what this score was prior to the implementation of the design, when the property was in agricultural production. The property also has high breeding bird and mammal diversity.

Additionally, a five-year evaluation of the plant responses to the 2001 removal of invasive shrubs and vines documented that by 2007, native plant species richness within upland habitats increased by nine species and the percentage cover of native plants increased by 15 percent. Additionally, native plant species richness in the lowland habitats of the EcoLab increased by three species and the percentage cover of native plants increased by 14 percent. Wetlands that had been spared the effects of invasion by Amur honeysuckle exhibited no changes in species richness or percent coverage of native plants during the same time period. Overall, these results highlight the positive impacts that the 2001 removal of invasive shrubs and vines had on native plants within the EcoLab.

Although much of Jensen's design focused on installation of plants within the Riverdale estate, it is notable that his manipulation of hydrology by digging ponds and wetlands and the removal of some of the subsurface drainage tiles may have been some of the most beneficial manipulations he made. Removal of these drainage tiles, plus the accidental clogging of others, led to the restoration of several springs and the reestablishment of a more natural hydrological regime within the EcoLab. Over a long time period, this altered hydrological regime led to the Clover Meadow changing from a plot of nonnative turf grasses and clover into a hardwood swamp. Breeding American woodcock (*Scolopax minor*) and beaver (*Castor canadensis*) have been documented in this location. The presence of beaver is likely another cause of the continued success of the restoration, because their presence has helped maintain a more natural hydrological regime.

The Marian College EcoLab (currently the Marian University Nina Mason Pulliam EcoLab) case study illustrates how Jensen's designed landscape laid the foundation for a modern-day ecological restoration project. At the time Jensen designed the Riverdale estate, plant availability was likely limited, the knowledge of ecology was elementary, and ecological restoration had not been conceptualized. Yet the project was successful in part because Jensen applied the knowledge of ecology and botany learned from Cowles and his other ecologist friends to design landscapes in a way that anticipates the practice of modern-day ecological restoration. In conclusion, we feel Jens Jensen himself would consider his historical efforts and the current Marian University restoration efforts to be important because the site provides refuge from modern life for the people of a large urban city and an educational center that promotes the beauty and value of natural habitats in the Midwest.

CONCLUSIONS

Our syntheses of the relationships among pioneers in ecology and landscape architecture and the history of the Marian University Nina Mason Pulliam EcoLab highlight how historical precursory restoration efforts lacking an ecocentric approach laid the foundations for ecological restoration in the Midwest. The relationships between these pioneers also highlight the strong historical link between ecology and ecological restoration in the region, and they bring to the forefront the contributions that Chicago- and Illinois-based individuals and institutions made toward the development of both fields. The history of the EcoLab illustrates how a precursory restoration effort lacking an ecocentric approach provided the foundation from which ecological restoration could emerge as the discipline of ecology matured. In conclusion, the lessons learned from these precursory restoration efforts highlight the shared heritage of the science of ecology and its application to ecological restoration, which in turn confirms the importance of integrating science and practice within ecological restoration in the Midwest.

CHAPTER TWO

MIDWESTERN THEORY AND PRACTICES THAT HAVE SHAPED THE FIELD OF ECOLOGICAL RESTORATION

CHRISTIAN LENHART, PETER C. SMILEY JR., AND JOHN SHUEY

INTRODUCTION

Ecological restoration is a goal-driven enterprise focused on assisting the recovery of damaged, degraded, and destroyed ecosystems (SER 2004). Practitioners perceive environmental degradation as an opportunity for recovery that can complement the broader goal of improving ecological integrity. In the process, ecological integrity is defined within the context of our overarching restoration goals and project resources. Site-based restoration goals may vary widely and include multiple objectives such as enhancing ecosystem services, biodiversity conservation, recreational opportunities, or even simple aesthetics. Often the goal is to reestablish ecological processes and communities that self-organize into functional, resilient ecosystems capable of adapting to changing conditions. The unifying attribute that defines ecological restoration relative to more practically oriented management endeavors (such as forestry or watershed management) is a reliance upon ecological principles to inform both restoration outcomes as well as development of restoration strategies. By infusing ecological theory into restoration strategies and our assessment of restoration outcomes, we ensure that we can commit the resources required to initiate the restoration project and follow through with the necessary management and maintenance necessary for its long-term success.

Too often, ecological restoration is misconstrued as primarily an art or skill (Van Diggelen et al. 2001) rather than the science-informed discipline that it is. Ecological restoration has long been criticized for lacking a broad theoretical framework to predict the outcome of restoration actions (Bradshaw 1987). In large part, this misconception is due to poor documentation of a priori goals and assumptions that many practitioners simply take for granted. For example, a casual observer of the restoration of Indiana oak savanna mosaics described in Chapter 7 may easily walk away with the impression that native plants were simply planted within former agricultural fields, when this restoration effort was much more complex. Restoration of these Indiana oak savanna mosaics incorporated climate change adaptations and attempted to address issues involving regional population dynamics of insect, amphibian, and reptile communities; the population genetics of local ecotypes; plant community patch dynamics; and downstream nutrient loading and water export. Overall, these restoration strategies were designed to preserve the evolutionary trajectories of the native flora and fauna residing within this dynamic ecosystem and were based on general and prevalent theories in ecology and conservation biology.

A major dilemma for the field of ecological restoration is that it is not rocket science; in other words, it cannot simply be reduced to the laws of physical science. We do not have an analog to Einstein's theory of general relativity that can be used to plot precise trajectories of future ecosystem succession and development. The term *theory* in ecological restoration is used more broadly than in other scientific fields and includes guiding principles having a mixture of physical science, ecology, and management precepts. While trajectories of ecosystem succession often move in predictable directions, they do not follow precise paths and are easily interrupted by disturbance, competition, predation, and anthropogenic impacts to the landscape. Trajectories of ecosystem change are sometimes obliterated by invasive species or catastrophic disturbance. In other words, natural ecological processes, in the form of disturbance events, often disrupt trajectories of ecosystem succession, sometimes to the point that it is easy to give up in frustration and think we can never fully comprehend the complexities and mechanics behind our restorations. Yet ecosystems usually behave in predictable manners that well-designed and executed restoration projects can exploit because the core foundation of our underlying theories and assumptions are sound. By incorporating ecological theory into our restoration efforts, they are guided toward resilient and dynamic ecological outcomes. Restoration outcomes that have the internal redundancy and elasticity required to react autonomously to change are likely to be the most successful outcomes. Indeed, ecological restoration is more complex than rocket science. Our underlying theories are not linear or deterministic, or likely to produce predictable, stable equilibria (Falk et al. 2006). Instead, our underlying theories are complex and illuminate the boundaries within which healthy ecosystems fluctuate in response to perturbations.

Many of these ecological theories have emerged from the Midwest, long attributed to be the birthplace of ecological restoration (see Chapter 1). In this chapter, our objective is to review selected theory and restoration practices that originated in the Midwest and discuss how these developments have shaped the field of ecological restoration.

ROLE OF THEORY IN SHAPING THE FIELD OF ECOLOGICAL RESTORATION AND VICE VERSA

Finding broad theoretical principles that apply to ecological restoration universally has been elusive. A number of factors have contributed to this problem that include (1) the unique nature of each ecosystem and region; (2) the practical nature of ecological restoration; and (3) the inability to conduct controlled, well-designed scientific studies at the ecosystem scale. Yet restoration practitioners have developed many new concepts that in turn have shaped the direction of academic research indirectly, driving the development of theory (see Chapter 3). A good example is the development of invasive species principles, which were driven by the experience of natural resource managers and restoration practitioners. Invasive species have been major problems for agriculture, forestry, fisheries, and other more utilitarian natural resource management fields for centuries. Farmers and natural resource managers have long tried to manage weeds such as Canada thistle (*Cirsium arvense*) and destructive animal pests such as brown rats (*Rattus norvegicus*) and common carp (*Cyprinus carpio*) in the Midwest (see Chapter 8).

It was not until later in the twentieth century and early twenty-first century that scientists began to identify principles of invasiveness. Instead of waiting for the next invasive species to arrive and wreak ecological havoc, scientists began to identify the organismal traits and environmental settings that would favor invasion. The ability to identify invasive species based on traits is a critical component of evaluating the risk of invasive species and effective prevention and control efforts (Van Kleunen et al. 2010). Early efforts at identifying traits of invasive species were descriptive narratives like Baker's (1974) list of traits of the ideal weed (Cadotte and Lovett-Doust 2001). Quantitative evaluations were conducted later by either comparing invasive and non-invasive species with trait information available within botanical compendiums or with experimental studies measuring traits of invasive and non-invasive species in field and laboratory settings. Baker (1986), in a review of North American plant invasions, identified that successful invaders display any combination of seven traits: (1) the climate in the place of origin is similar to the colonizing habitat; (2) the invading species have a similar life-form to the native taxa; (3) the soils of the place of origin are similar to the colonizing habitat; (4) the invading taxa exhibit generalized pollination via wind or insect vectors or a self-pollination system; (5) the invading taxa possess dispersal systems that enable them to be mobile in new habitats; (6) the breeding systems of the invasive species allow for sexual reproduction at low densities without inbreeding depression and are able to promote genetic variation through recombination; and (7) the invasions into dense communities where native taxa rely on vegetative reproduction are facilitated by invaders having vegetative reproduction. A recent meta-analysis of 117 studies involving 125 invasive plant species (Van Kleunen et al. 2010) found that invasive species differed from non-invasive species in physiological, leaf-area allocation, shoot allocation, growth rate, size, and fitness traits. Van Kleunen et al. (2010) also observed that invasive species often possess multiple correlated traits that favor fast growth, and thus the challenge is determining which traits directly confer invasiveness and which traits are simply correlated with the trait that confers invasiveness.

Such management-oriented theories of invasiveness may not be universally applicable since plant establishment, competition, and succession may occur differently in older, unglaciated landscapes compared to young, recently glaciated landscapes typical of the Midwest. Although falling short of universality, such management-oriented theories provide the ability to predict and subsequently anticipate which invasive species may be problematic in certain settings. These types of theories have been helpful in improving restoration practitioners' ability to develop and implement successful restoration projects, even if they must extrapolate the relevance to a specific site or region.

Another way in which ecological theories have shaped the field of ecological restoration is through the development of criteria for success and failure. Having quantifiable objectives is critical for objectively assessing the success or failure of restoration projects (Zedler 2007). It is straightforward to identify specific success criteria, such as a maximum allowable percentage of bare ground or the percentage of coverage by grasses versus forbs in a prairie restoration. While specific criteria are based on experience from past projects done in a specific setting, the chosen criteria for evaluation is often dictated by larger underlying goals that are shaped by our view of what success means. For example, goals for species composition and/or diversity are often dictated by what type of plant community or ecosystem was thought to exist prior to major changes from human interventions. In this situation, restoration success would be evaluated based on the degree of similarity in species composition and diversity between the restored ecosystem and a reference ecosystem.

Many recent restoration projects have adopted a more functional approach to evaluating restoration success. While prairie restoration efforts often adopt plant species diversity and/or establishment of rare species as a common goal, more recently restoration goals have become more functional by specifying coverage by certain plant guilds or functional groups (e.g., C_3 and C_4 grasses). In the past decade, carbon storage for climate change mitigation has become an increasingly common goal for terrestrial and wetland restoration projects. Wetlands in particular can be large carbon sinks, so work in this area is only likely to expand in the future (see Chapter 4). The provision of creating pollinator habitat has become more and more important in prairie projects as well with the decline in native bee (clade Anthophila) populations. Numerous initiatives to promote pollinator habitat on roadsides, in native landscaping, and in riparian habitats now exist across the Midwest.

In other areas, such as wetland mitigation, for example, Midwest states have developed precise guidelines to ensure successful restoration from a legal standpoint, since mitigation wetlands are legally required "replacements" for wetlands filled or impacted by development (Matthews and Endress 2008). Wetland mitigation rules tend to focus on the area of wetland established and the hydrologic conditions needed to support wetland vegetation, defined as hydrophytic plants by the U.S. Fish and Wildlife Service. There are also usually regulations that require specific levels of native plant coverage, which typically are more modest than the targeted levels of native plant coverage used by an ecological restoration project that is intended to achieve the levels identified by reference ecosystems.

Jordan et al. (1987) originally thought monitoring of restoration projects would be a way practitioners could contribute to the science and theory of restoration, because it was envisioned that practice would serve as a vehicle for collecting data and learning about the ecosystems being restored. However, monitoring of restoration projects by practitioners has not been done widely in a systematic way that would promote learning across the Midwest. The vast majority of small, private restoration projects do not integrate research or require monitoring simply because funding for monitoring and evaluation is not provided (although some federal and state programs require monitoring to be included as part of the project design; for example, see the Minnesota Wetland Conservation Act). Another reason past monitoring efforts of restoration projects have not contributed to the development of theory is because they are regulatory in nature, as in the case of mitigation wetlands. Flawed experimental designs have also hindered objective statistical analysis, which better enable people to extrapolate their results beyond the individual project site. Consequently, the lessons learned from post-restoration monitoring are often not transferrable to other regions or ecosystem types, thus limiting their contribution to the development of more broadly applicable scientific theories.

RELEVANT ECOLOGICAL THEORIES ORIGINATING FROM THE MIDWEST

Succession

Succession is one of the oldest theories in ecology, and it is highly applicable to the science and practice of ecological restoration. After all, what is the traditional process of restoration but the manipulation of the physical and biological characteristics of a degraded ecosystem to alter the species composition of a community so that it resembles a pristine community? Succession also serves as the foundation for conceptual models that predict how degraded ecosystems change through time following restoration (Suding and Gross 2006; Hobbs and Suding 2009).

Early Midwest ecologists were instrumental in developing the foundations of succession theory. Henry Chandler Cowles's field research on plants in the Indiana Dunes (see Chapter 1) provided the first documentation of complete successional seres (Real and Brown 1991). Cowles assumed that changes in plant communities between locations within the Indiana Dunes represented the community changes that occurred through time. Cowles's observations and research findings were important because they highlighted that plant communities were not static entities. Although Cowles recognized successional seres and that succession drifted toward a stable equilibrium, he did not believe the equilibrium state could be achieved (Real and Brown 1991). Cowles envisioned succession as a never-ending process of nonlinear change (Real and Brown 1991).

Frederic Clements (1916), as a result of his research in Minnesota and Nebraska, developed a theory of succession that postulated that changes in plant communities over time occur through an orderly, directional, and predictable process. Clements also analogized successional development of plant communities with the ontogenetic development of organisms, and referred to communities as superorganisms. Initial seres were thought to modify the physical environment to enable other species to establish within the community. Clements predicted that succession would result in the development of recognizable seres in an expected order until a balance between biotic and abiotic conditions was achieved. At this point, the plant community would reach a stable endpoint called the "climax community," which was considered to be a self-perpetuating community best-adapted to the climate of a given area. In the absence of disturbance that would reset a community to an earlier sere, succession was predicted to lead to the establishment of the climax community across broad climatically defined regions (Clements 1936).

Henry Gleason (1917) developed a contrasting framework of succession called the individualistic concept of plant ecology. Based on his field experience in Illinois and Michigan, he noted that Clements's concept of succession assumed too much homogeneity. Gleason argued that identifiable seres were similar to one another only in degrees and that upon close examination, these community types were not real or natural units. Gleason considered each species an independent entity and proposed that its distribution was dependent upon its unique evolutionary and ecological heritage. Gleason postulated that chance played a large role in the development of plant communities and plant communities did not follow a predictable trajectory to a specific community type. Gleason's work was largely ignored until the 1950s, when John T. Curtis at the University of Wisconsin–Madison (see Chapter 3) and Robert H. Whittaker at the University of Illinois independently evaluated whether plant species responded individually to changes in environmental conditions (Waller et al. 2012). Their research findings, particularly Curtis's landmark book *The Vegetation of Wisconsin* (1959), are credited with expanding Gleason's theory into the widely recognized continuum concept that postulated that species composition within communities varies continuously along environmental gradients (McIntosh 1995; Waller et al. 2012).

Raymond Lindeman's research on plants and animals within Cedar Bog Lake, Minnesota, extended the theory of succession from one focused on describing changes in species composition of communities through time to one describing changes in the ecosystems through time in terms of energy flow through the ecosystem (Real and Brown 1991). The extension of succession to ecosystems is the basis of more recent conceptual models (Suding and Gross 2006; Hobbs and Suding 2009) describing the trajectory of degraded ecosystems after restoration. Lindeman's classic paper (1942) is also considered the first successful holistic ecosystem analysis and was instrumental in the development of ecosystem ecology (McIntosh 1981), which is relevant to modern-day restoration with its increasing focus on restoring ecosystem function.

Succession theory has undergone many developments since the early contributions of Midwest pioneers in ecology and now exists as a hierarchically structured theory with multiple propositions, a corresponding law, and individual models describing how the law applies to specific situations (Pickett et al. 2011a). The law of succession states that community structure will change through time as a result of disturbance, differential species availability (i.e., colonization, existing seed banks, survivors), and differential species performance (i.e., physiology, life history, facilitation, competition, etc.) (Pickett et al. 2011a). Current succession theory encompasses processes occurring within different organizational levels (i.e., individuals, populations, communities, ecosystems) at multiple spatial and temporal scales. The most important change relevant to ecological restoration is that the theory is now capable of accounting for different responses under a wide variety of conditions. The ability to account for site-specific differences will increase the relevance of succession theory to the field of ecological restoration and in turn may enable the practice of ecological restoration to contribute to further developments of the theory.

Succession theory also led to the development of important concepts related to the nature of communities and ecosystems, and whether equilibrium exists within ecological entities. Classical equilibrium perspective within ecology viewed the community as a self-organizing entity that would develop predictably to a final stage that represents the equilibrium state. Early equilibrium concepts were in part based on the underlying assumption that divine intervention promoted order and stability in nature (Botkin 1990). Stephen Forbes's paper (1887) on the lake as a microcosm is one of the earliest discussions of how natural selection promotes the balance of nature within populations and communities in floodplain lakes in Illinois. Clements's concept of the community as a superorganism that developed into a stable climax community is another good example of the classical equilibrium perspective. Much early ecological theory in the twentieth century involving population and community ecology (e.g., Lotka-Volterra predator-prey dynamics, the logistic growth curve and maximum sustainable yield) incorporates equilibrium and the balance-of-nature concept, and these theories viewed populations and communities as structured, regulated, steady-state entities unless disturbed by humans (Botkin 1990).

Although equilibrium-based theories in ecology have been challenged since the 1930s (Botkin 1990), they remain influential even though modern-day ecologists more readily recognize the role of random forces in structuring populations, communities, and ecosystems. The traditional concept of ecological restoration—the return to a former self-organizing state—is indicative of the influence of the balance-of-nature and superorganism concepts on the field of ecological restoration. Likely those who question the feasibility of restoring damaged ecosystems are influenced by the balance-of-nature concept that views humans as destructive forces whose intervention interferes with nature's self-organizing capacity. Non-equilibrium concepts favored in modern-day ecology have led to recent changes in the concept of ecological restoration. If equilibrium does not exist in nature, then it is not feasible to create self-organizing ecosystems. Indeed, current definitions of restoration appear to incorporate non-equilibrium concepts through the greater emphasis on (1) recovery of degraded ecosystems rather than replicating a specific community or ecosystem type; (2) restoring ecosystems that exhibit trajectories different from the degradative trajectories that occurred before restoration; and (3) restoring ecosystem function over ecosystem structure (Clewell 2009).

Relationships of Biological Diversity with Ecosystem Diversity and Stability

Pioneering research in the Midwest has contributed to an understanding of the relationships of biological diversity with ecosystem diversity and stability. These relationships are fundamental to ecological restoration because many restoration projects attempt to increase ecosystem diversity in an attempt to improve biological diversity. Additionally, if increased biological diversity conveys increased ecosystem stability, then the potential for restoration success increases.

The relationships of biological diversity with ecosystem diversity and stability are critical assumptions underlying all restoration projects. Specifically, it is assumed that increasing ecosystem diversity (i.e., physical habitat diversity) will result in increased biological diversity, which in turn leads to increased ecosystem stability. The first documentation of the relationship between biological diversity and habitat diversity in streams was described in Gorman and Karr's landmark publication in 1978. Gorman and Karr sampled fishes and measured water depth, velocity, and substrate types in streams in Indiana and Panama, and they found that fish diversity was positively correlated with habitat diversity there. Additional stream fish research conducted in the Midwest by Issac Schlosser, James Karr, and Karr's students led to the development of a model of stream fish communities that describes how these communities change with habitat heterogeneity and pool development (Schlosser 1987; Smiley and Gillespie 2010). In turn, this research evaluating the relationships between stream fish communities and habitat conditions within channelized and unchannelized streams in the region led to the development of the Index of Biological Integrity that is used widely throughout the United States to evaluate water quality (Smiley and Gillespie 2010).

In 1982, David Tilman began a long-term project to examine the relationship between ecological stability and botanical diversity in grasslands located at Cedar Creek Natural History Area, just north of Saint Paul, Minnesota. Tilman's research experimentally delineated the relationship between diversity and stability in plant communities (Tilman and Downing 1994) and documented that a strong positive correlation existed between plant diversity and plant community. Tilman and Downing argued that biological diversity increases stability at the community level because the differential species' responses to disturbance or stress cumulatively produce stable community dynamics through time. Communities with low species diversity are likely to respond to stress with fluctuating biomass production, while increased species diversity increases community stability. As a consequence, restoration practitioners often aim for diverse communities to increase ecosystem stability as well as to reduce the threat of invasive species, thus reducing follow-up management needs.

Concepts Developed in Response to Lake Eutrophication

Ecological and limnological theory related to aquatic ecosystems developed in parallel with theories based on terrestrial ecosystems in the 1900s, and later would provide guiding principles for lake and river restoration. Lakes served as an early laboratory for the development of ecological and limnological principles because the lake ecosystem is visibly contained within discrete boundaries at a scale that is possible to quantify. Consequently, scientists were able to identify many important physical and biotic processes that influence lake communities and ecosystems. Eutrophication is one of the foremost problems facing freshwater and marine ecosystems today. The problem has stimulated research within lakes to understand the process and to evaluate methods of controlling it (Cooke et al. 2005; Schindler 2006). Scientists from the Midwest have contributed significantly to development of concepts related to lake eutrophication.

University of Wisconsin–Madison faculty member Arthur D. Hasler was among the first to call attention to the negative impacts of cultural eutrophication of lakes through inputs of domestic sewage and agricultural land use, as well as the difficulty of restoring eutrophic lakes (Hasler 1947). Hasler also pioneered the use of manipulative whole lake experiments as a way of increasing the understanding of lakes and guiding lake management

(Johnson and Hasler 1954; Hasler 1964). Whole lake experiments have been instrumental in increasing our understanding of eutrophication. For example, whole lake experiments conducted in Canada resolved the limiting nutrient controversy in the 1970s and established firmly that phosphorus was the primary factor causing eutrophication within lakes (Schindler 2006).

The challenge of restoring eutrophic lakes also stimulated the development of two related concepts within lakes—trophic cascades and alternative stable states—that in turn contributed to the development and evaluation of an important lake restoration practice: biomanipulation. Stephen R. Carpenter (University of Notre Dame), James Kitchell (University of Wisconsin–Madison), and James R. Hogson (Saint Norbert College) hypothesized that trophic cascades could explain annual variances in lake trophic state (Carpenter et al. 1985). Specifically, they proposed that piscivory suppresses planktivorous fish, which increases abundance of zooplankton. Increased zooplankton abundance subsequently leads to reduced algal abundance. Subsequent research by Stephen Carpenter and his University of Wisconsin–Madison colleagues that involved whole lake experiments, small-scale enclosure experiments, and paleolimnological studies in experimental lakes in Wisconsin (Carpenter and Kitchell 1996) further increased the understanding of trophic cascades and the importance of biotic interactions in determining lake trophic state. Additionally, Carpenter et al.'s (1985) trophic cascade hypothesis is considered one of the most significant concepts in modern limnology (Cooke et al. 2005).

Scheffer et al. (1993) proposed that shallow lakes prone to algal blooms could exist in two alternative stable states. One is a clear-water state dominated by aquatic plants, and the second consists of an algal-dominated state that is less biologically diverse and less attractive for recreational activities. Transitions between these two states can be caused by trophic cascades, increased nutrient loading, or factors that cause declines of aquatic plants (Scheffer et al. 1993; Dent et al. 2002). This concept is an example of nonlinear equilibrium models that have been used to explain changes in ecosystem states following disturbances within lakes and rivers (Dent et al. 2002). Although the shallow lakes alternative stable states concept was developed based on field observations from Europe and Australasia (Scheffer et al. 1993), it and other examples of alternative stable states in lakes have been applied to lake management and restoration in the Midwest (Carpenter et al. 1999; Dent et al. 2002; Hobbs et al. 2012; Chapter 4).

RESTORATION PRACTICES PIONEERED IN THE MIDWEST

Use of Reference Ecosystems as Restoration Targets

A number of restoration practices used nationally and internationally have been developed in the Midwest. The use of native plants in landscaping, particularly in public parks, was an important precursor to the field of ecological restoration (see Chapter 1). People needed to gain an appreciation for the value of individual plant species before they would value restoration of whole native plant communities or ecosystems. Jens Jensen and others associated with the prairie style of landscape design promoted landscape design projects based on the composition and structure of Midwest ecosystems (see Chapter 1). The use of native ecosystems as a basis for landscape design was a precursor to the current restoration practice of using reference ecosystems (high-quality ecosystems) to develop restoration targets.

Closely related to our increased awareness of the value of native plant species was the effort to inventory native plant communities in different Midwest states while relatively pristine plant community remnants could still be surveyed before they became impacted by expanding development, agriculture, and other impacts in the region (Sears 1925; Curtis 1959). As people began to restore prairies, they found they needed guidance on which species to plant. Early restoration efforts used nearby remnants (i.e., reference ecosystems) as a guide for their restoration projects (see Chapter 3). Later efforts used natural history books such as *An Annotated Flora of the Chicago Region* (Pepoon 1927) and *The Vegetation of Wisconsin* (Curtis 1959) to determine which species would be expected to occur in different ecosystems (see Chapter 3; Stevens 1995). The reference ecosystem concept has served as a driving principle in many restoration efforts in the Midwest, particularly in prairie and upland restoration projects where this goal appeared to be reasonable if not entirely attainable (see Chapter 3). The reference ecosystem concept crosses restoration specialties as it is used in both terrestrial and aquatic restoration projects. For example, the natural channel design approach to restoring streams and rivers (see Chapters 5 and 9) uses geomorphic and hydrologic information from nearby sites as the basis for restoration design.

The scientific and practical merit of the reference ecosystem concept has become a source of much recent debate in ecological restoration. One such debate involves the novel ecosystem concept (Hobbs et al. 2009), which postulates that it is impossible or at least not beneficial to attempt to restore plant communities or whole ecosystems to their historic ecosystem structure and function. The novel ecosystem concept consists of a framework describing the degree of ecosystem alteration ranging from minor vegetation shifts to complete physical alteration. Restoration of historic conditions might be possible in degraded ecosystems that have experienced minor changes. In contrast, restoration of historic conditions would not be possible in severely degraded ecosystems (e.g., mine quarries), and restoration efforts would only produce ecosystems containing novel species combinations.

The novel ecosystem concept has likely influenced restoration in the Midwest less than other parts of the world because it is still possible to restore the plant composition of a Midwest prairie to a resemblance of its pre-1900s condition if the physical environment has not been substantially altered. In regions that have undergone intensive mining and agriculture, such as western Australia, where Hobbs is based, it can be nearly impossible to restore anything resembling historical ecosystem structure and function. Additionally, the fertile Midwest prairie soils are also amenable to seeding and rapid plant establishment unlike the old, infertile soils covering much of Australia and other old, low-fertility landscapes. Many Midwest prairie and savanna restorations still strongly rely on historical reference information as a guide (Egan and Howell 2001). However, more and more functional approaches to restoration in the region are being promoted, especially in highly modified urban and agricultural landscapes (see Chapters 6 and 9).

Oak Savanna Restoration Techniques

The story of the development of oak savanna restoration techniques represents the practice of restoration as a way of learning about the ecology of native ecosystems. Oak savannas are ecosystems for which there were no remaining high-quality remnants that could serve as a guide for developing restoration targets. On the other hand, the original scientific thought regarding oak savanna ecosystems was that they were not unique ecosystems but simply prairies with trees (Stevens 1995). In the late 1970s, Stephen Packard and a small group of volunteers began attempting to restore prairies in Chicago-area forest preserves along the North Branch of the Chicago River that were suffering from fire suppression and invasive species (Stevens 1995). These early attempts at prairie restoration were small-scale experiments in which Packard and the volunteers tested different techniques and then noted the ecosystem responses (Stevens 1995). At the time, a guidebook for restoring prairies was not available, so Packard and his crew had to learn by doing. They experimented with manual brush removal and the use of fire followed by planting prairie seeds. Their early attempts at prairie restoration within oak groves were not successful, as evidenced by increasing amounts of thistle and briars growing under the oaks instead of prairie plants.

Packard, in considering this problem, read up on species accounts of the plants and concluded that instead of planting prairie plants, they should be planting savanna species in the oak understory to restore oak savannas (Stevens 1995). Packard compiled a list of potential savanna species and used it to develop a seed mix consisting of half savanna and half prairie species. The savanna/prairie seed mix was planted in shaded areas in 1985, and by the spring of 1986, many of these savanna species had emerged instead of the thistle and briars that characterized these areas in previous planting attempts (Stevens 1995). The work by Packard and the North Branch volunteers furthered the understanding of oak savanna ecology and restoration by identifying indicator species unique to oak savannas and using low-intensity prescribed burns to remove invasive species. Notably, neither Packard nor the volunteers were academically trained ecologists or scientists.

While the Packard and North Branch volunteers' restoration efforts were successful, in 1996 their efforts raised enormous controversy among urban and suburban residents who objected to the removal of large trees, use of herbicides, and deer removal (Gobster 2000). The Chicago Restoration Controversy, as it became known, stimulated much reflection about the social and cultural aspects of ecological restoration (Gobster 2000). The work of Packard and the North Branch volunteers represented the birth of Chicago Wilderness, one of the largest volunteer restoration organizations in the Midwest.

Prescribed Fire

The use of prescribed fire in prairie restoration was developed primarily in the Midwest. Native Americans were thought to have started fires to keep areas more open for hunting and walking. With European settlement in the

Midwest in the late 1700s to mid-1800s and the displacement of the Native Americans, fire ceased as a functional process on the landscape. It was not until wildlife managers such as Aldo Leopold in the 1930s recognized the value of fire for keeping grasslands open for game bird species that fire was recognized again as a management tool (Leopold 1933). The University of Wisconsin–Madison Arboretum experimented with prescribed fire as part of prairie restoration in the 1940s, and based on the outcome of their research findings, they began implementing it regularly in 1950 (see Chapter 3). Eventually the use of this practice spread to other smaller prairies on private and public land. The early guides for prescribed burning were written by practitioners. For example, Dane County, Wisconsin, naturalist Wayne R. Pauly wrote a highly referenced guide in 1985 that was intended to provide novices with information about how to conduct small prairie fires with handheld equipment and a minimum number of inexperienced assistants. The study of fire effects on prairie and forest ecosystems has since spread and is an area of active research in the Midwest. More recently, practitioners have explored the combination of grazing, burning, and mowing to maintain prairies and savanna understories (Helzer 2009).

Phosphorus Control in Lakes

Lake restoration has developed separately from the field of ecological restoration. Limnologists view lake restoration in the United States as a young discipline that began in the 1970s, and they use the term *lake restoration* to refer to the reestablishment of important missing or altered processes, habitats, concentrations, and species (Cooke et al. 2005). Lake restoration has focused primarily on the problems of eutrophication or acidification, not whole ecosystem restoration (National Research Council 1992; Cooke et al. 2005). As a result, lake restoration goals and objectives differ considerably from those used as part of restoration projects in other ecosystems. The goal for lakes is not based on the structure and function of reference lakes, although some aspects of ecosystem structure may be included, such as the reestablishment of historically important plants like wild rice or wild celery.

Early attempts to address eutrophication in the 1960s involved the use of chemical or mechanical in-lake practices to reduce algae (Cooke 2007). Controlling phosphorus by reducing external input via physical and chemical methods was the primary strategy for reducing eutrophication in lakes in the 1970s (Shapiro et al. 1975). For example, the cleanup of Lake Erie in the 1970s following its "death" focused on removal of phosphorus from known point sources (pipe outlets) and in laundry detergent (Ashworth 1986). These efforts improved water quality in Lake Erie and greatly reduced the occurrence and extent of algae blooms in the western part of Lake Erie (Makarewicz and Bertram 1991). Widely considered a great environmental success story by 1990, the algae blooms reemerged in the 2000s with increasing intensity (Kane et al. 2014). Today, scientists, managers, and lake restorationists are focusing on control of agricultural non-point pollution to reduce the dissolved phosphorus that is thought to be the cause of the recent algal blooms (Michalak et al. 2013). Lake Erie demonstrates the difficulty of restoring a large body of water when most of the watershed is in row-crop agricultural land use.

In addition to watershed management, many in-lake practices were developed to reverse eutrophication through phosphorus control, invasive aquatic plant removal, and other chemical, mechanical, and biological means (Cooke et al. 2005). The use of alum to bind and precipitate phosphorus on the lake bottom has been employed in many eutrophic Midwest lakes, particularly in urban areas (Cooke et al. 2005). In lakes with a lot of recreational use, there is demand to accelerate the removal of phosphorus stored in the lake sediments to make the lakes more amenable for boating and fishing. A more sustainable in-lake restoration practice is biomanipulation (Shapiro et al. 1975), which originally was considered to consist of a range of practices involving the manipulation of lake biota and habitats intended to reduce algal biomass. Recently, the term *biomanipulation* has been more narrowly defined as practices that lead to reductions of the abundance of small planktivorous fish by increasing the density and amount of piscivorous fish, which enables increases in herbivorous plankton that consume algae (Lathrop et al. 2002; Cooke et al. 2005). Biomanipulation was pioneered in Minnesota by Joseph Shapiro as an economical alternative to traditional chemical and engineering methods of reducing eutrophication (Shapiro et al. 1975; Shapiro and Wright 1984). The feasibility of biomanipulation was then further evaluated by research on the trophic cascade concept conducted by University of Wisconsin–Madison scientists (Carpenter et al. 1985; Carpenter and Kitchell 1996). Short-term evaluations of biomanipulation indicate it is capable of reducing eutrophication in shallow eutrophic lakes (Lathrop et al. 2002; Schindler 2006). Long-term evaluations of biomanipulation on lakes in Wisconsin and Minnesota also confirm it is capable of inducing the clear-water state within eutrophic lakes (Lathrop et al. 2002; Hobbs et al. 2012), although in some cases it might be temporary. These two long-term evaluations also suggest that in many cases biomanipulation is a practice that should be used in conjunction with non-point pollution control in the lake watershed and with practices that alter internal nutrient cycling (Lathrop et al. 2002; Hobbs et al. 2012).

Two-Stage Channel Design

Lotic ecosystems are inherently more dynamic than lakes and terrestrial ecosystems. It has long been recognized that streams are dynamic in both the variability of water conditions at one point in time (depth, velocity, temperature) and in their movement and changes to dimensions over time (Leopold et al. 1964). Streams are driven by physical forces (flowing water and sediment movement) exhibiting irregular episodic events more than biological forces in comparison to terrestrial ecosystems. Therefore, concepts from the physical sciences, especially geomorphology and hydrology, have been used to guide stream and river restoration. Notably, stream and river restoration has been influenced strongly by concepts from the field of fluvial geomorphology because these ecosystems are shaped by flowing water and sediment transport. Channel evolution models were developed by geomorphologists to describe and predict changes to stream and river dimensions and physical characteristics over time following disturbances such as channelization or increases in discharge (Schumm 1979, 1981). Channel evolution models have been instrumental in developing strategies for stream restoration by helping to diagnose underlying drivers of channel change. Specifically, these models have been used to select locations to target restoration efforts within channelized streams and to design restoration assessment efforts (Shields et al. 1998).

In the Midwest, the desire to restore stream ecosystem functions led to new approaches in the design and management of agricultural drainage ditches (i.e., channelized agricultural headwater streams). Channel evolution models are less applicable for designing stream restoration projects within agricultural watersheds in the Midwest because of the practice of channel maintenance, which regularly reshapes the channel via dredging. The high degree of physical alteration found within agricultural drainage ditches combined with the cultural need for agricultural drainage makes it difficult to restore these degraded streams to anything close to a reference condition. Restoration efforts that focus on reestablishment of selected ecosystem functions and that maintain the ability of these streams to provide agricultural drainage will more likely be widely adopted by the agricultural community. Most agricultural drainage ditches in the region are locked in place by channel maintenance performed regularly to maintain a straight, trapezoidal form that prohibits the natural processes of lateral migration and point-bar building. With these logistical challenges in mind, the alternative drainage design called the two-stage channel design was developed and is promoted as a restoration practice in the Midwest, particularly Ohio and Indiana (Powell et al. 2007a, 2007b; NRCS 2007).

The traditional design for channelized headwater streams is an overly large trapezoidal cross section capable of holding a 100-year flood within its stream banks (NRCS 2007). The two-stage channel design involves altering the cross section of the trapezoidal channel by widening the top banks and establishing benches intended to function as miniature floodplains within the channel (NRCS 2007). The two-stage design is essentially the channel-within-a-channel design that has been used as an alternative design in channelized streams since the 1970s (Brookes 1988; Landwehr and Rhoads 2003). The application of the channel-within-a-channel design to agricultural drainage ditches in the Midwest was pioneered in Ohio and Indiana in the early 2000s (Powell et al. 2007b). The design is only appropriate for use in low-gradient channelized headwater streams that are not undergoing incision (NRCS 2007). Potential benefits of the two-stage channel design include reduced channel maintenance as a result of increased downstream transport of fine sediment, reduction of nutrient transport, and improved aquatic habitat (Powell et al. 2007a, NRCS 2007).

Current evaluations of the two-stage channel design within the Midwest indicate that it (1) provides limited reductions in nitrate export, which varies among streams and discharge levels (Roley et al. 2012; Mahl et al. 2015); (2) leads to highly variable reductions in turbidity (Mahl et al. 2015); (3) may not promote organic matter breakdown (Griffiths et al. 2012); and (4) may not increase fish and aquatic macroinvertebrate biodiversity (Janssen 2008). However, previous studies within an Illinois channelized stream with naturally formed benches suggest the two-stage channel design may provide hydraulic refugia for aquatic animals during flood events (Schwartz and Herricks 2005) and increases hydraulic diversity (Rhoads et al. 2003; Rhoads and Massey 2012). It should be noted that previous research efforts evaluated the impact of the two-stage channel design by itself. Highly altered ecosystems such as channelized streams likely require combinations of restoration and watershed management practices to address the broader impacts of agriculture (see Chapter 9).

Dam Removal

While the two-stage channel design can be viewed as making the best of a bad situation from an ecological perspective, the removal of aging and nonfunctional dams is viewed as one of the best ways to restore processes in altered river environments (Bednarek 2001). Instead of trying to re-create the former ecosystem structure, dam removal attempts to restore ecosystem function by removing barriers to its operation. Removing dams reestablishes the hydrologic and sediment transport regime as the free flow of water is restored and fine sediments are mobilized with the formation of a new channel within the sediments. The passage for fish and other aquatic life is reestablished, allowing almost instantaneous upstream movement for fish and other mobile aquatic animals. Despite the benefits of dam removal, there are many challenges associated with these projects (see Chapter 5). The mobilization of sediment stored upstream of the dams is a major issue, as these sediments may contain contaminants and their release following dam removal may impact downstream benthic organisms, such as freshwater mussels (family Unionidae, family Sphaeriidae).

Dam removal in the United States has been conducted since 1915, although the number of dams removed has increased dramatically in the past two decades (Bellmore et al. 2017; Connor et al. 2015; Service 2011). It is estimated that by 2020 at least 80 percent of the two million dams that exist in the United States will be greater than fifty years old (Bellmore et al. 2017). The increasing number of aging dams suggests that frequency of dam removal will likely continue to increase in the future. The Midwest has been identified as a leader with respect to dam removal and the subsequent evaluation of these dam removals (Doyle et al. 2005; Bellmore et al. 2017). Particularly, Wisconsin has been identified as a leader in dam removal (Service 2011; Bellmore et al. 2017). Wisconsin's success has been attributed to a state grant program that provides a 50 percent cost share for dam repair and removal and the advocacy of well-organized groups within the state promoting dam removal (Pohl 2002).

Nationally and regionally within the Midwest, the three primary reasons that dams are removed are for environmental, safety, and economic reasons (Pohl 2002). In the 1970s and 1980s, safety was the primary reason for dam removal, but beginning in the 1990s, environmental concerns became the primary reason (Pohl 2002). Particularly within Wisconsin and Minnesota, many small mill and water storage dams have been removed as part of stream and river restoration efforts (Pohl 2002).

CONCLUSIONS

Pioneers in ecology and ecological restoration from the Midwest developed important concepts and practices that influenced the field of ecological restoration in the Midwest and internationally. The key lesson is that these concepts and practices evolved through time, reinforcing the dynamic nature of ecological restoration. Future ecological restoration efforts in the Midwest will face a number of challenges such as climate change (see Chapter 7), invasive species (see Chapter 8), and increasing urbanization (see Chapter 6) and agricultural land use (see Chapter 9). The future of restoration in the region will depend on the contributions of scientists and practitioners working together with governmental and nongovernmental organizations to ensure that the science and practice of ecological restoration continues to evolve to become more effective at repairing damaged ecosystems.

PART

TWO

Case Studies—

Building on Lessons

from the Past to

Forge a New Future

CHAPTER THREE

HISTORICAL AND CURRENT PRAIRIE RESTORATION IN THE MIDWEST

STEVE GLASS AND DARYL SMITH

INTRODUCTION

Ecological restoration as a distinct professional discipline dates only from the 1980s with the formation of the Society for Ecological Restoration (see Chapter 1). However, its roots stem from many fields, ranging from traditional ecological knowledge to sustainable resource management. Restorations have been attempted in one form or another for many years (see Chapter 1). From early on, the act of restoring the land has been an interdisciplinary effort (Howell et al. 2012). It was perhaps George Perkins Marsh, in his classic *Man and Nature*, who made the first explicit calls for the restoration of lands damaged by human activities when he urged that forests be replanted and restored (Lowenthal 1965). Many of today's restorations had their origins in projects conceived of and designed by garden writers, landscape architects, and horticulturists in Europe and America who were active between 1870 to 1940 (Howell et al. 2012; see Chapter 1).

Prairie restoration was one of the earliest forms of ecological restoration in the Midwest, dating from the mid-1930s (Jordan and Lubick 2011). Two of the earliest prairie restoration projects in the Midwest were initiated in 1935 and were about ninety-six kilometers apart. One project was located at the University of Wisconsin–Madison Arboretum (UW Arboretum) and the other at Aldo Leopold's family farm near Baraboo, Wisconsin. These two early prairie restoration efforts, which survive to this day, laid down a rich legacy that inspired and informed the vast network of people involved in Midwest prairie restoration through the years and contributed to today's global ecological restoration movement.

There are several likely reasons why prairies were among the first ecosystems in the Midwest to be the subject of restoration efforts. First, the prairie, the predominant plant community type in the western part of the Midwest, was rapidly disappearing. Under pressure from the plow and cow, the prairies were slipping away. By the 1930s, large, continuous, undisturbed tracts of prairie no longer existed. But visions of this once-extensive prairie landscape were within the living memory of some, and these people, noticing what they had lost, were eager to do something to retrieve it. Secondly, prairie plants, and prairies themselves, are on a human scale and are easily appreciated. A prairie can be planted with simple and common agricultural tools and methods. These tools—the plow, the disk, the rake, and shovel—were readily available in the agricultural community. Thirdly, it was thought at the time that planting a prairie would be a relatively easy project. Aldo Leopold's instructions to Theodore Sperry, who supervised the early plantings at the UW Arboretum, were *"go plant a prairie."* Despite the simplicity of his initial charge, Leopold anticipated it might require fifty years for a mature prairie to be established. The theory was that a prairie could be planted in a season and would mature in a human lifetime, unlike a forest, which contains trees that dwarf a person and is a plant community that can take decades, if not centuries, to mature. Finally, while the prairie was mostly gone from the landscape, scattered prairie remnants remained that were capable of serving as models or templates to guide planting efforts. Remnants provided not only a historic reference but also supplied seeds and plants for the early prairie restoration efforts. For example, the earliest plantings in Curtis Prairie relied upon the outdated and destructive method of using prairie hay, plants, and prairie sod taken from remnants in southwestern Wisconsin.

The idea of prairie restoration caught on. In the late 1940s, the UW Arboretum undertook a second effort at Greene Prairie, which was located few kilometers south of Curtis Prairie. Then came prairie plantings at Green Oaks (Knox College) in Galesburg, Illinois, followed by other Illinois prairie restorations at Fermilab and Morton Arboretum. Since then, prairie restoration has become a phenomenon, spreading like wildfire across the Midwest. There are now likely thousands, if not tens of thousands, of prairie plantings, re-creations, reconstructions, and restoration projects across the region.

The popularity of prairie restoration among individuals, organizations, schools, institutions, colleges, universities, and businesses is probably explained by the fact that it is an accessible activity. It is often done on weekends by community work parties or individuals in their own backyard in their spare time. Although prairie seed can be expensive to purchase, individuals and organizations can often supply their own seeds because it is relatively easy to collect, process, and store prairie seed. Seedling plants are now readily available and represent a luxury that the pioneers of prairie restoration did not have. It is also relatively easy to prepare the site and to plant, especially since it does not require vast amounts of land to start a prairie planting. Prairie restoration can be done on a range of spatial scales from the backyard, the schoolyard, highway roadsides, and rights-of-way to large spatial scales spanning thousands of hectares. Notable large-scale prairie restoration projects include the Nachusa Grasslands and the Midewin National Tallgrass Prairie in Illinois and at the Neal Smith National Wildlife Refuge in Iowa.

Prairie restoration is done for many different purposes. Originally, prairie restorations like the UW Arboretum's Curtis and Greene Prairies and Aldo Leopold's Prairie in Baraboo, Wisconsin, were attempts to include all attributes of historic natural communities (composition, structure, function, processes, service, and aesthetics) to the extent possible. This is the most idealized and most difficult ecological restoration goal. Now, prairie restorations are also done for a variety of other purposes that include (1) aesthetics, as typified by the small projects commonly conducted at arboreta and botanical gardens; (2) to provide ecosystem services or improve ecosystem function, such as those projects completed at landfills; (3) to control invasive plants, as popularized in Iowa by the Integrated Roadside Vegetation Management Program; (4) for educational purposes, like prairies restored on school campuses; and (5) to promote civic engagement and ecological literacy.

Our objectives in this chapter are to discuss in detail two case studies of Midwest prairie restoration projects to give the reader a sense of the history of the science and practice of ecological restoration, its midwestern roots, and today's diversity and scope of restoration goals and project types. The two projects we profile are interesting and complementary case studies.

One, Curtis Prairie, represents the historical and ecocentric restoration ideal, while Iowa's Integrated Roadside Vegetation Management Program represents restoration for the anthropocentric objective of controlling weeds on roadsides.

INTRODUCTION TO CURTIS PRAIRIE

The work of the scientists and practitioners involved with Curtis Prairie was literally and figuratively groundbreaking and laid down a fertile path that resulted in the emergence of not only one of the first ecological restoration projects in the world but also the discipline of ecological restoration. This achievement places them among the pioneers of the discipline and makes their story an important chapter in the history of ecological restoration. The story of Curtis Prairie is significant for the field of ecological restoration from two different perspectives. First, it is a journey of self-discovery and a tale of the evolution of the project's purpose and potential. Secondly, the Curtis Prairie restoration developed fundamental theoretical and practical principles of ecological restoration in general and prairie restoration in particular. Curtis Prairie from the start has stimulated practical and applied research—today what we would call adaptive management—on topics such as fire, germination requirements of prairie species and establishment techniques, composition, the functioning of prairies in Wisconsin, and changes over time in prairie community structure. Thus, this initial restoration effort was a pilot project for the link between theory and practice.

Curtis Prairie is named after John T. Curtis, who was the UW Arboretum's first director of plant research and a University of Wisconsin Department of Botany faculty member (Sachse 1974). Curtis served as director of plant research alongside Aldo Leopold, who was the director of animal research (Sachse 1974). With the exception of the time period from 1942 to 1945, Curtis served the university from 1937 until his death in 1961. He was a leader in ecology whose work contributed to the development of the continuum theory (see Chapter 2) and resulted in the development of statistical methods that were precursors to ordination methods widely used in ecology today (Waller et al. 2012). He is best remembered for actively applying the ecological principles he was developing to practical conservation problems (Howell and Stearns 1993). Curtis wrote the first master development plan for the UW Arboretum ecological communities (Kline 1993). He and his students conducted field experiments at the UW Arboretum that led to the development of prairie restoration methods. They also conducted detailed studies of plant communities throughout Wisconsin that resulted in *The Vegetation of Wisconsin* (Curtis 1959). This book is a classic in ecology (Fralish et al. 1993) that contributed to the development of ecology theory and serves to guide restoration efforts through its documentation of historic plant species composition and dynamics.

The first prairie plantings at Curtis Prairie were undertaken for the sake of the plant community itself, in an effort to restore the species and complexity of the historic communities and to re-create a sample of original Wisconsin to study. It is important to note that these early projects were not known as ecological restoration, as that term did not come into use until the early 1980s, when it was coined by William R. Jordan III and Kurt Wendt at the UW Arboretum. The prairie plantings at Curtis Prairie were bold undertakings because at the time, there was only a limited amount of information about the variety of prairie types across the state of Wisconsin and the species composition, structure, and function of prairie ecosystems. Furthermore, there was no information on how to restore a prairie from scratch. In fact, the process was almost taken for granted—as Leopold simply stated, "*Go plant a prairie*," as if it were obvious how one would do so. As Leopold expected, the Curtis Prairie project turned out to be a great object for study by providing insights into the value of restoration practice and ecological theory, and the importance of ecological restoration as a form of environmental management and a technique for research (Jordan et al. 1987). Our objectives for this section are to (1) summarize the experiences and accomplishments of those involved with Curtis Prairie; (2) identify the links between theory and practice that occurred as a result of this important restoration effort; and (3) discuss how the efforts at Curtis Prairie contributed to the development of other prairie restoration projects in the Midwest.

Getting Started

The idea of restoring a prairie was first proposed by University of Wisconsin–Madison professor Norman Fassett, a taxonomist and botanist. The idea of restoring a prairie was to support Aldo Leopold's vision of re-creating on the property a sample of what Dane County, Wisconsin, looked like when the early settlers arrived (Sachse 1974; Kline 1993). Fassett's goal was to re-create samples of prairie to be used for teaching and research that would identify areas where further research was necessary (Cottam 1987). Fassett's reference ecosystem for the first plantings were the remnant patches of native prairie in Dane County in southwestern Wisconsin and those remnants located on the high bluffs in the western part of the state along the Mississippi River. These native prairie remnants also served as source materials (seeds, sod, and transplants) for the early plantings. A group of faculty members assisted by graduate students did the planting. The University of Wisconsin–Madison did not fund the project, and there was little involvement from the general university or the community of greater Madison. The team chose sites that were slated for destruction because of highway and building construction or farmland expansion. The distinction of making first plantings at Curtis Prairie in 1934 (Sachse 1974) belongs to Fassett and his student John Thomson, who later would become an accomplished University of Wisconsin–Madison professor of botany in his own right.

Curtis Prairie was planted on 0.24 square kilometers of horse pasture that was previously prairie. The site of the future prairie was dominated by annual bluegrass (*Poa annua*), Kentucky bluegrass (*Poa pratensis*), and other ruderal species such as dandelion (*Taraxacum* spp.) (Cottam 1987), which persist to this day. The team selected the site because it was available, had what was thought to be good soils from an agricultural perspective, and it was close to headquarters and the workforce. The site was also close to existing roads and footpaths, so little was required to establish the needed transportation infrastructure (Sachse 1974).

Continued Planting Efforts

Theodore M. Sperry, a newly minted PhD of Arthur Vestal's from the University of Illinois, arrived in Madison, Wisconsin, in the winter of 1936 to take on his first big job—supervising the new prairie planting project at the UW Arboretum.

As mentioned earlier, Sperry's recruiter and boss Aldo Leopold told him simply to "*go plant a prairie.*" This mandate sounded simple but proved to be anything but. At the time Sperry began work, this kind of task had not been done before, and he would be among the first anywhere to attempt a restoration at this scale. Leopold's directive raised a number of practical and theoretical questions that ecologists were only beginning to ask and did not have the answers for. These were important questions, such as Can a new prairie be planted? How does one go about planting a prairie? Should one use seeds or plants or prairie sod or prairie hay? How are prairies managed and maintained? What is a prairie? What species is it composed of? Are there different kinds of prairie? How do prairies function? These and other questions emerged gradually and in sequence as the prairie-planting project got under way.

Just prior to Sperry's arrival, the UW Arboretum received a source of labor from the federal government. From 1935 to 1939, a Civilian Conservation Corps (CCC) camp was based at the UW Arboretum. The CCC was a public works project enacted by President Franklin Delano Roosevelt during the Great Depression to put the unemployed to work (Meine 1988). Almost overnight, the CCC became the greatest infusion of labor the environmental and conservation movements have ever seen (Meine 1988). The 300 young unemployed men stationed at Camp Madison were put to work at the UW Arboretum. They continued the plantings and experiments in Curtis Prairie under Sperry's supervision in addition to the other work they did at the arboretum. Funds for housing and wages for the CCC enrollees were provided by the federal government. It is likely that without the CCC, the UW Arboretum would not have undertaken restoration work.

Planting efforts were discontinued during World War II and resumed around 1950 under the direction of Grant Cottam, one of Curtis's graduate students, and David Archbald, the first managing director of the UW Arboretum (Sachse 1974). The new planting efforts involved the use of seed mixtures and various planting methods such as burning before seeding, disking after seeding, and mulching with prairie hay (Cottam and Wilson 1966).

Experimentation

From the start, prairie restoration has been a combination of the practice of restoration—the doing—and scientific inquiry—the development and testing of theory. At its best, the practice and science of ecological restoration is a seamless interaction, with practice informing science and science in turn developing theories for practice to test. This mutually beneficial interaction between science and practice was a wonderfully serendipitous result of the UW Arboretum experience.

Historically, Curtis Prairie was not set up as a grand experiment, mainly because no one knew enough at the time about prairie ecology or restoration or all the variables involved in conducting a rigorous prairie planting experiment (Cottam 1987). Nonetheless, experimentation at a more practical level was involved from the beginning.

Early planting efforts in Curtis Prairie focused on finding the best planting method. Researchers experimented with various techniques such as removing surface soil, plowing, seeding, transplanting sod, and mulching with prairie hay to transform the horse pasture into prairie. Instead of adopting a pure trial-and-error approach, Fassett, Sperry, and Leopold made sure that every step in the initial prairie restoration project was turned into a research project. For example, Fassett and Thomson set up small experimental plots for the first major plantings of forty-six species. The early work by Fassett, Thomson, and Sperry set the tone for later work, and Archbald carried out additional plantings and experimentation between 1950 and 1957 (Sachse 1974).

This research and experimentation led to an early decision about the best way to continue to establish the prairie. Curtis judged that although transplanting prairie sod dug from nearby remnants was the most effective method, it was too expensive and too labor intensive for the large scale they had in mind (Howell and Stearns 1993). He determined that seeding was more affordable—a finding that foretold the widespread use of seed mixes in future prairie restoration efforts. Curtis and Henry C. Greene also began experiments on the germination requirements of prairie plant seeds. They discovered that most prairie species require cold stratification for germination (Greene and Curtis 1950).

The most well-known example of this mutually beneficial interaction between science and practice was the realization that prescribed fire was a necessary prairie management tool. Researchers and land managers tending Curtis Prairie noticed that Kentucky bluegrass and other invasive species persisted despite the best efforts to eliminate them. The difficulty in establishing seedlings and in discouraging Kentucky bluegrass led Curtis to study the problem and search for solutions. Curtis and Partch (1948)

FIGURE 3. Curtis Prairie after a prescribed management burn.

conducted a series of studies of fire effects on prairie species and the plant community itself to test the notion that fire could control these invasive plants. These initial results suggested that prescribed fire could do the trick (Curtis and Partch 1948), and from that, prescribed fire became a routine management practice at the UW Arboretum that continues to the present (Figure 3).

Fire is prescribed in two senses. In one sense, it is a management recommendation to cure what is perceived by managers to be what ails the prairie. In a second sense, it is an application of fire conducted under strictly preestablished, or "prescribed," conditions. These prescribed conditions include minimum and maximum wind speeds, temperature, humidity, as well as a specified wind direction and other factors related to safety. The Curtis Prairie fire research findings also confirmed that prairies are fire-dependent ecosystems—they are ecosystems that require fire for survival. These research results further demonstrated that humans could reintroduce the natural process of fire into restorations and remnant plant communities. As a result, prescribed fire has become an important restoration and management strategy and technique that is routinely used in fire-dependent communities in the Midwest and around the world. Other innovations that came from the early prairie restoration projects at the UW Arboretum included the practice of seed collection, processing, and storage, and eventually, the idea that seed mixes and ecotypes could be tailored for various site conditions and community types.

Questions Generated by the Restoration Process

Kline and Howell (1987) observed that prairies are especially suitable for certain kinds of research and for answering questions that arose as a result of experience with tallgrass prairie restoration in southern Wisconsin. Specifically, prairie restoration is suitable for addressing the following research questions: Why are certain prairie species underrepresented? What explains the presence and abundance of weedy introduced herbs? Why do woody species invade prairies, and how may they be controlled? (Kline and Howell 1987). Kline and Howell also felt that questions related to long-term community dynamics (i.e., disturbances, population explosions, local ecotypes, changes in forb frequencies) could be examined during the prairie restoration process.

In addition to the identification of the importance of prescribed fire for prairie restoration and health (Curtis and Partch 1948; Howell and Stearns 1993), the act of prairie restoration at the UW Arboretum led to the discovery of allelopathy in showy sunflower (*Helianthus laetiflorus*). Curtis and Cottam (1950) noticed that extensive patches of native showy sunflower excluded other species and that the patches, while viable on the perimeter, were hollow in the center. Their research on this observation demonstrated that this native sunflower was allelopathic (poisoned other species) and autotoxic (inhibited its own growth).

The act of prairie restoration also led to the refinement of the use of prescribed fire at the UW Arboretum to control invasive species. UW Arboretum ecologist Virginia Kline (1985) conducted a series of fire studies that demonstrated that the every-other-year prescribed fire schedule promoted—rather than discouraged—the continuation of the pest species white sweet clover (*Meliotus alba*) and yellow sweet clover (*Melilotus officinalis*). Based upon the study's recommendations, the UW Arboretum fire crews began to use an early spring burn one year to stimulate synchronous germination of white and yellow sweet clover seeds followed by a late spring burn the following year to kill developing second-year individuals of these two species.

The process of re-creating a prairie also demonstrated the lack of knowledge of the species composition of prairies, how they functioned, and the variety of prairie community types. This knowledge gap was the inspiration for the detailed field studies conducted by Curtis and his students of remnant plant communities in Wisconsin that resulted in the groundbreaking book *The Vegetation of Wisconsin* (Curtis 1959).

Community Dynamics and Monitoring

The planting of the Curtis and Greene Prairies provided a wonderful opportunity to observe the ways in which the species composition of prairies changed over time (Cottam and Wilson 1966; Sperry 1984). Introduction of plants in Curtis Prairie set the stage for a long-term study of movement of species within the developing community (Cottam 1987). Blewett (1981) performed an exhaustive analysis of the results of five plant surveys of the Curtis and Greene Prairies conducted between 1952 and 1976. The results of this analysis indicated that species composition within these restored prairies changed over time as the planted species established themselves in optimum sites and as the species responded to short-term climatic events prior to the surveys (Blewett 1981; Cottam 1987).

Sperry apparently conducted the first monitoring survey of the original plantings within a portion of Curtis Prairie. Although Sperry's planting work at Curtis Prairie ended in 1941 when the Civilian Conservation Corps camp closed, he made three return trips, in 1946, 1982, and 1990, to conduct monitoring and to evaluate the success of the project. Sperry (1984) found that 55 percent of the original 237 plantings of 46 species were successful and 45 percent unsuccessful. Sperry (1984) also noted that several successful species spread widely over the prairie, while other well-established species exhibited little or no spread between 1941 and 1982. Based on these findings, Sperry concluded that the Curtis Prairie restoration project was successful.

Management Plans

At the time the Curtis Prairie project was initiated, virtually nothing was known about restoring prairies. There were no planting guidebooks to serve as references and no species lists to consult. There were no local restoration experts and no native plant nurseries from which to purchase plants. These early restorationists had no precedents and no restoration examples to use as a guide, and they had to discover the prairie restoration process as they went along. In the beginning, restorationists often worked without a formal written plan. The absence of a formal plan does not mean that this restoration project lacked planning. In fact, the early restorationists were excellent ecologists, landscape architects, and record keepers who brought extensive field experience and then modern ecological theory to the task.

Sperry, in particular, left behind a rich and detailed record of his work, including observations, planting records, and analysis of the results. In notebook after notebook, in tight script and precise prose, he tells where and when each species was collected—seed, sod, or plant—and where, when, and how it was planted. His notebooks provide documentation about the weather on the planting day, whether or not the species was watered, and a general estimation of its survival rate by the end of the planting season. Sperry's notebooks also contain supplemental accounts of the operation of the nursery, special experiments, photography, the labor costs, observations of species growth and survival, and field maps. Sperry and the other early restorationists used what we now call an adaptive management approach, carefully documented their work, and built research experiments into the project.

In 1951, Curtis wrote the first conceptual plan for the UW Arboretum, which was basically a general layout map of plant communities. In 1992, Kline wrote an updated version of the Curtis plan (Kline 1992) that contemporized the UW Arboretum's vision, reviewed the status of the Curtis Prairie and Greene Prairie restoration projects, and recommended what management actions were still needed. Kline's updated plan is essentially a master plan that establishes the use policies for research, management, and visitors at the UW Arboretum and, similar to the Curtis plan, lays out the boundaries of the different plant communities on the UW Arboretum grounds.

Looking to the Future of Curtis Prairie

The generic restoration goals and objectives for Curtis Prairie (i.e., *go plant a prairie*) have had their advantages and disadvantages. The current goal for Curtis Prairie is to develop a good example of tallgrass prairie containing the species appropriate for the moisture continuum that exists on site (Kline 1992). The vagueness or perhaps inability of outreach specialists to adequately explain the purpose and nature of the desired outcomes has encouraged generations of stakeholders, researchers, managers, faculty, staff, students, volunteers, and the general public to stamp the project with their own views of what is desirable and possible. This freedom to interpret the Curtis Prairie project has been beneficial because it has encouraged, in the social realm, a good deal of public interest and involvement—a citizenry of diverse voices and opinions. On the other hand, this freedom of interpretation has created the challenging task of reconciling the competing viewpoints of what is socially desirable with the reality of what is ecologically attainable.

Lack of precision in goals and objectives has created difficulty in selecting the variables to measure and monitor, and in determining whether the desired outcomes have been obtained. Nonetheless, an assessment of long-term restoration success in Curtis Prairie has recently been compiled (Wegner et al. 2008). The assessment asked and answered four basic questions: Is Curtis Prairie uniformly rich in species? Is Curtis Prairie restored? Does Curtis Prairie match natural prairies in species richness? What are the persistent restoration issues?

The 2002 vegetation survey revealed that Curtis Prairie had 265 species, 230 of which are native (Snyder 2004). Diversity varied across the prairie, with the greatest diversity in the unplowed remnant and lower diversity in wetland areas dominated by the invasive reed canary grass (*Phalaris ardundinacea*) or woody native pest species such as gray dogwood (*Cornus racemosa*) and sandbar willow (*Salix interior*).

Any restoration is a process rather than a single event because the possibility exists that the restoration process will never be completed. Curtis Prairie is no exception. For example, the site is too small to provide habitat for some species of native grassland birds that would be expected in midwestern prairies. Likewise, there are some native mammal species that have not yet colonized the site. One notable success is that a breeding pair of sandhill cranes (*Grus canadensis*) has nested in the prairie for the past several years.

Comparing the species richness of Curtis Prairie with natural prairies is a difficult issue to evaluate because there are so few remaining native prairie remnants with similar soil and other site conditions, comparable management histories, and existing data sets. Yet, when looking at the entire prairie ecosystem and remnants within Wisconsin, Curtis Prairie has less species richness than some of the better prairie remnants, which have up to 400 or more species.

Native and nonnative woody and herbaceous pest species continue to be widespread despite seventy-five years of active management (McGaw 2002; Snyder 2004). The density and frequency of these invasive species reduce native species diversity, suppress flammable fuel loads, and make prescribed fires difficult. In some cases, the early lack of understanding of the use of fire and cutting resulted in stimulating the spread of nonnative species via their root systems. The reduction of aspen (*Populus tremuloides*) was attempted in Curtis Prairie during the early years but resulted in its greater spread, which has yet to be brought under full control.

Stormwater runoff has a tremendous impact on Curtis Prairie. It is estimated that 64,141 cubic meters of stormwater enters Curtis Prairie annually. The stormwater introduces excessive amounts of sediment, pollutants, and seeds of pest species into Curtis Prairie; facilitates the spread of reed canary grass; and has eroded a ditch in the center of the prairie.

Smoke management issues and public safety precautions to protect dense commercial and residential developments in the urban setting require that prescribed fires be conducted under relatively safe and cool conditions. These conditions might reduce the effectiveness of the prescribed fires in controlling pest species as compared to the effectiveness of naturally occurring fires in the past. Despite these issues, Curtis Prairie has been an active and engaging player in the promotion, practice, and study of ecological restoration and restoration ecology through its historical and current involvement in prairie restoration. Generations of faculty, staff, and students have contributed to restoration studies at the UW Arboretum, and their work has generated hundreds of research papers leading to the development of many restoration innovations and research methods. Current research at Curtis Prairie takes advantage of grant opportunities, management problems, and faculty and graduate student research interests, and it relies upon an adaptive restoration framework. Ongoing studies focus on linking research and land care by developing adaptive management solutions to urban restoration challenges such as stormwater runoff and invasive species management.

Prairie restoration involving restoration of remnants with relict species and reconstructing prairies from scratch have been ongoing activities in Iowa since the mid-1960s. A program introduced in the mid-1980s to plant prairie species along roadsides to assist with weed control has proven to be a valuable form of prairie restoration across the state. Invasive plants were the primary constituents of the roadside vegetation of Iowa state and county roads in the 1950s, and they created logistical challenges for the management of these areas. A perspective of the history of the development of roads and associated changes in roadside vegetation is helpful in understanding how the ascendancy of invasive species occurred in a landscape dominated by prairie at the time of settlement in 1832.

Iowa was 80 percent prairie in 1832. As the state was being settled, the first roads followed existing trails into the prairie wilderness. Initially, there was little disturbance of the native vegetation adjacent to the roadways, so most of the early roadsides consisted of native prairie. Most disturbances of that roadside prairie came from incidental road widening from increased traffic and minor modifications to drain wet spots. Early construction impacts were moderate, as roadside ditches were constructed with hand tools and horse-drawn scoops. Often, construction did not impact the entire right-of-way, leaving patches of prairie. Usually any prairie vegetation removed from the roadsides was soon replaced by secondary succession with seed from nearby prairie. Advances in construction machinery resulted in more earthmoving and greater amounts of prairie remnants being eliminated from the roadsides. Prairie reestablishment via secondary succession was still possible as long as original prairie was nearby and/or remnant patches remained in the rights-of-way. Meanwhile, expansion of agriculture in the late nineteenth and early twentieth century converted much of the remaining prairie to cropland and eliminated remnant prairies adjacent to roadsides. Consequently, opportunities for secondary prairie succession in rights-of-way declined (Smith 1995).

The final elimination of roadside prairie remnants in Iowa began in the 1920s with a program to get Iowa motorists out of the mud. This program created a statewide improved farm-to-market road network and eliminated most roadside prairie remnants. With more modern construction equipment, newly constructed roads were extensively reshaped, creating a firmer, raised roadbed and more defined roadside ditches. The rights-of-way were considerably modified, as soil had to be moved to form the raised roadbed and deepen the ditches for better drainage. With more extensive earth movement, features like ditch foreslopes and backslopes became common and prairie vegetation less common. Additionally, most of the adjacent land is now agricultural cropland, and the remaining native vegetation was limited to isolated prairie remnants and fencerows. No nearby remnants remained to provide seed for secondary prairie succession (Thompson 1989; Smith 1995).

The loss of remnant seed sources coupled with increased soil disturbance by road construction and erosion from adjacent cropland created ideal conditions for invasive species. Unless the roadsides were intentionally seeded, they readily became prime sites for early successional non-prairie species. Early roadside seeding mixtures consisted largely of nonnative grasses smooth brome (*Bromus inermis*) and occasionally fescue (*Festuca* spp.). Nonnative prairie species and native woody species easily invaded nonnative grass plantings on certain soil types. The invading species were a visual blight and raised concerns regarding possible invasion into adjacent cropland (Cramer 1991).

Mowing was used to control weeds in the roadsides. Initially, local farmers mowed the roadside ditches adjacent to their property. Eventually, extensive mowing programs were developed as county and state agencies assumed increasing responsibility for roadside maintenance. Often, rights-of-way were mowed fence to fence to improve the appearance of the roadside. This weed control technique was costly because it was labor intensive and required a heavy investment of time and considerable equipment maintenance. As herbicides became more readily available, they were used as an alternative weed control method to reduce the demand on the labor force. County agencies usually contracted with private companies to broadcast spray the roadsides under their jurisdiction with herbicides effective in the control of broadleaf and woody species. Herbicide application was less labor inten-sive, but it eventually became quite expensive as herbicide and application costs increased (Smith 1995). Eventually, many Iowa counties could not afford the high cost of spraying their many kilometers of secondary roads. Yet neither mowing nor herbicide spraying was effective in weed control. Both techniques tended to stress the perennial vegetation and open sites for invasive species (Christiansen and Lyons 1975). In addition, the use of large quantities of herbicide in broadcast spraying increased the potential for groundwater contamination. By the middle of the 1980s, it was readily apparent that a more effective, cost-efficient means of roadside vegetation management was needed.

Development of the Integrated Roadside Vegetation Management Program

The Integrated Roadside Vegetation Management Program was developed by two Iowa counties, Black Hawk and Story, as a viable alternative for agencies seeking a cost-efficient, ecologically based means of roadside weed control. This program combines the use of native prairie plantings with limited application of selective weed control techniques to increase the effectiveness and cost efficiency of roadside vegetation management. The use of mowing is greatly reduced, and spot spraying of herbicides replaces broadcast spraying. The increased awareness that 99.9 percent of the state's original prairie had been converted to cropland and the ineffectiveness of existing roadside vegetation management programs were major contributors to the development of the Iowa Integrated Roadside Vegetation Management Program.

Studies of native prairie roadside plantings in the 1960s and early 1970s laid the ecological groundwork for this innovative program in Iowa (Landers et al. 1970; Christiansen and Lyons 1975). The Iowa landscape at the time of settlement was dominated by tallgrass prairie (80 percent), and prairie openings were also present among oak-hickory (*Quercus* spp.- *Carya* spp.) savanna areas. The most stable, diverse plant communities are those native communities that have naturally adapted to a particular region over time (Odum 1971). Consequently, the Iowa Integrated Roadside Vegetation Management Program is based on the establishment of native prairie vegetation along the roadsides using plant species best adapted to the area. The stable, diverse native prairie plantings tend to maintain themselves and resist weedy invasion (Weaver 1954). Natural resistance to weedy invasion reduces the need for mowing and chemical control. In addition, the diversity and adaptability of native prairie sustains the roadside vegetation through adverse conditions (Figure 4).

Establishing prairie along roadsides is a utilitarian form of prairie restoration. The steep slopes of roadside ditches and compacted subsoils from

FIGURE 4. Gravel county road in Iowa with roadside prairie planting.

construction are challenges for seeding and plant establishment. However, more than 360 square kilometers of roadsides were planted with prairie species between 1988 and 2016. The extensive fibrous root systems of prairie plants assist with increasing plant survival following exposure to drought conditions and high salt concentrations often found along roadsides. These root systems also enable native prairie plants to outcompete the more shallow-rooted invasive weeds for water, nutrients, and space to grow (Christiansen and Lyons 1975). Many native prairie species are adapted to low nitrogen levels and tend to do well in poor soil. Often, prairie species will establish and grow in B horizon (subsoil) or C horizon (substrata), while non-prairie or invasive species cannot.

The benefits of restoring prairie along roadsides make the extra effort worthwhile. Grasses and other wildflowers add a rich aesthetic quality and present motorists with a wide variety of forms, textures, colors, and hues that change with the season. The diverse prairie communities provide valuable habitat for birds, small mammals, pollinators, and other wildlife species (Adams 1984; Camp and Best 1994; Delaplane and Mayer 2000; Forman et al. 2003; Smith et al. 2010), contributing thus to the conservation of native fauna as well as the flora. These restored prairies also serve as dispersal corridors for connecting widely separated prairie patches. Primary roads with wider rights-of-way function better as prairie corridors, as roadside prairies on secondary roads often occur as a narrow strip with an extensive edge that allows invasive species easy access and exposes the fauna to predators.

Restoring prairies along the roadsides provides functional benefits as well as aesthetic and ecological benefits. Wide swaths of prairie vegetation along roadsides can increase the storage of snow in ditches and reduce the amount of snow on the road. The deep, fibrous roots of prairie plants are effective in holding soil, reducing erosion, and stabilizing ditch slopes (Weaver 1954). Restoring prairie along roadsides increases the interception and infiltration of rainwater. Prairie plants are capable of intercepting and holding a considerable portion of rainwater because the surface area of their foliage is five to twenty times greater than the soil surface beneath it (Weaver 1954). For example, 0.01 square kilometers of big bluestem (*Andropogon geradii*) can intercept approximately 119 metric tons of rainwater during a 2.54-centimeter rainfall event (Clark 1937). The extensive root systems of prairie plants increase the soil's ability to take up and hold rainwater via infiltration. These roots can penetrate two meters or more into the soil and create air pockets and channels within the soil. The roots also provide large quantities of organic matter to the soil (Dierks 2011). Soil organic matter has the ability to hold up to 90 percent of its weight in water and promotes clumping and aggregate formation that increases soil porosity (Funderberg 2001). Increased water infiltration, increased soil organic matter, and stable soil aggregate formation stimulated by the roots of prairie plants can in turn reduce soil erosion. Universal soil loss equation calculations predict that increasing soil organic matter from 1 to 3 percent can reduce erosion by 20 to 33 percent (Funderberg 2001). Restored prairies can exhibit a rapid rate of increase of large water-stable soil aggregates, which is an indicator of increased soil organic content and soil quality. Levels of soil aggregates within a restored prairie on the grounds of Fermilab (Fermi National Accelerator Laboratory) in Illinois increased from 39 percent to 93 percent by the fifth growing season after planting and attained levels exhibited by nearby prairie remnants by the eighth growing season (Miller and Jastrow 1986).

The variability in roadside topography resulting from the presence of roadside ditches requires techniques for establishing and maintaining the prairie vegetation that differ from other types of prairie restoration efforts. New road construction often leads to heavily compacted soils that require special site preparation prior to seeding. Most county road rights-of-way are twenty meters wide with narrow, deep ditches. It is most effective to design native roadside seed mixes to include species adapted to a wide range of site conditions, from gravelly, well-drained soils at the top of the ditch to the heavy, water-saturated soils at the bottom of the ditch. The same seed mix can be seeded over the entire area and allow the species to find their niche. Wider rights-of-way more typical of primary highways may contain wet or dry areas of sufficient size to justify designing and planting a seed mix specific for these wider sites (Brandt et al. 2011).

Roadside seeding can be done with a native-grass drill, broadcast seeder, or hydro-seeder. Seeding on steeper slopes is usually done with a hydro-seeder. Seeding rates are determined by degree of slope and, to a lesser extent, seeding method. A typical prairie seed mixture on level sites consists of approximately 432 seeds per square meter with a 1:1 ratio of grasses (family Poaceae) to forbs. Steeper slopes require heavier seeding rates to allow for seed loss and plant establishment to control erosion. Recommendations for slopes are as follows: (1) 3:1 slope – increase by 50 percent to 648 seeds per square meter; and (2) 2:1 slope – increase by 100 percent to 864 seeds per square meter. Native grass seed is less expensive than native forb seed. Therefore, it is less costly to increase the amount of grass seed in the mix to achieve erosion control. However, at least a 25 percent forb component is needed for adequate diversity and long-term soil stability. When hydro-seeding with seed mixed in the slurry (one-pass method), increase seeding rates by 15 to 30 percent to compensate for seed hung up in the mulch (Brandt et al. 2011).

The Integrated Roadside Vegetation Management Program developed at a particularly appropriate time. Counties in Iowa were beginning to feel the pinch of the high costs of mowing and broadcast spraying of herbicides. County supervisors and staff were very receptive to a sustainable, ecologically sound program that was cost effective. The use of native prairie plants as part of the Integrated Roadside Vegetation Management Program provides value through the protection of groundwater with ecologically sound management practices, reduction of costs, and stabilization of roadside slopes with self-sustaining, locally adapted species. Currently, the Iowa Integrated Roadside Vegetation Management Program has restored native prairie vegetation to more than 360 square kilometers of Iowa roadsides, which is more than three times the amount of area (< 110 square kilometers) occupied by native prairie remnants in the state. Iowa, a state that originally contained more than 113,312 square kilometers of pre-settlement prairie within the heart of the tallgrass prairie region was ideally suited for this revolutionary project. Furthermore, there was a certain romantic appeal to recapturing a portion of our biological heritage. Three decades of success of this program are a testament to its viability and continuity.

CONCLUSIONS

The Curtis Prairie project is the foundation of prairie restoration. It provided the basic research and techniques for the development of the Integrated Roadside Vegetation Management Program and other restoration projects. The Integrated Roadside Vegetation Management Program in Iowa is a more utilitarian approach to prairie restoration that uses native prairie plants for weed control and the reduction of herbicide usage. However, this novel program, in contrast to the Curtis Prairie project, does not attempt to replicate the original prairie. Both approaches to prairie restoration are important because more natural areas are likely to be lost in the future, creating a need for restoration to maintain and increase the number of prairie ecosystems within the Midwest (Smith 2014).

The Curtis Prairie project not only restored prairies but it also generated knowledge. Thus, the narrative of Curtis Prairie is significant because it is a journey of self-discovery and a tale of the evolution of the project's purpose and potential. The Curtis Prairie narrative is also important because the UW Arboretum prairie restoration efforts led to the development of fundamental theoretical and practical principles of ecological restoration in general and prairie restoration in particular. The Curtis Prairie restoration has from the start stimulated practical and applied research on topics such as fire, germination requirements and establishment techniques for prairie species, the composition and function of prairies in Wisconsin, and temporal changes in prairie community structure. As such, the Curtis Prairie project was a pilot project for the link between theory and practice in ecological restoration.

The success of the Integrated Roadside Vegetation Management Program suggests that this utilitarian approach to prairie restoration may be a model for other restoration efforts with utilitarian objectives. Helzer (2012) opined that we need to restore the viability of the fragmented prairie landscape with the species and processes that enable the ecosystem to function and flourish rather than attempt to re-create historical prairie community structure. While not replicating the diversity of pre-settlement prairie, such utilitarian prairie restoration projects provide elements of that ecosystem. Projects such as the Curtis Prairie project will be essential in the future for continuing the development of fundamental theoretical and practical principles of ecological restoration.

CHAPTER FOUR

FLOODPLAIN WETLAND RESTORATION ALONG THE ILLINOIS RIVER

MICHAEL J. LEMKE, HEATH M. HAGY,

ANDREW F. CASPER, AND HUA CHEN

Rivers are dynamic ecosystems composed of the river channel and the wetlands and shallow lakes within the adjacent floodplains. Historically, the main driver of wetland development and succession within the floodplain was the timing, duration, and magnitude of inundation from river flood-waters. We define floodplain wetlands broadly as a continuum of aquatic areas located within the 100-year large river floodplain and include shallow lakes, wetlands, and surrounding areas that are periodically flooded with surface water.

The characteristics of floodplain wetlands and their biotic communities can vary substantially depending on the nature of the seasonal flood pulse (Junk et al. 1989; Ward et al. 2001). While the flood-pulse concept has provided a general theoretical framework for understanding river floodplain ecosystems, lessons learned from wetland restoration projects are needed to advance the science and guide future wetland restoration in large river systems. The management of many rivers in the Midwest has largely focused on the individual parts of the river (i.e., river channel, floodplain, dammed navigation pools) and has been segregated by traditional means of classification (i.e., lentic and lotic), despite substantial modifications to rivers in the twentieth century. Thus, our understanding of the entire large river ecosystem comes from the assembly of isolated parts rather than the interconnected whole. Additionally, further complicating our understanding of the entire large river ecosystem is that structure and function of the interconnected whole has been extensively modified by humans.

Floodplain wetlands in the Midwest have been lost or significantly degraded over the last century as a result of sedimentation, altered hydrology, and land-use conversion (Bellrose et al. 1979; Dahl 1990). In the Midwest, networks of levees were erected in the early twentieth century that isolated many floodplains from rivers to allow for development and agriculture. River levels have been raised and stabilized by locks and dams to facilitate commercial navigation, recreation, and other uses (Bellrose et al. 1983). The lock and dam systems have created a series of reservoirs (i.e., navigation pools) encompassing the main river channel and the former adjacent floodplains. These reservoirs permanently inundated many floodplain wetlands that formerly had seasonal or temporary water regimes. Late in the last century, commercial navigation generated more than $253 billion annually along the Upper Mississippi River, making a return to more naturalized flows unlikely in most areas (Sparks 1995; Sparks et al. 1998). These highly modified river systems have hydrologic regimes that preclude a simple approach of removing river levees to restore ecosystem services to river floodplains. The classic theoretical model of the river flood pulse as the driver of the form and function in floodplains should be reevaluated in these highly altered ecosystems (Junk et al. 1989; Jackson and Pringle 2010).

Despite extensive river modifications and loss of floodplain wetlands in the Midwest, there are a few examples of floodplain wetland restoration projects that can contribute to the science behind floodplain ecology and guide future restoration efforts (Figure 5). Evaluation of biotic responses (e.g., patterns of floral and faunal succession) to the manipulation of floodplain drivers (e.g., water inundation) will help hone restoration ecology theory and document efforts for more efficient conservation outcomes. While documented as one of the richest, most productive systems in North America, the Illinois River and its floodplains have been extensively degraded by navigation and agriculture that occurred between the 1930s and the 1990s. However, since the 1990s, conservation policy and restoration actions have shown that the ecosystem is resilient as well as able to support continentally significant biodiversity and productivity (Bellrose et al. 1983; McClelland et al. 2012). In this chapter, our objective is to summarize notable floodplain wetland and associated shallow lake restoration projects along the Illinois River and the lessons learned from each project to better guide future restoration efforts.

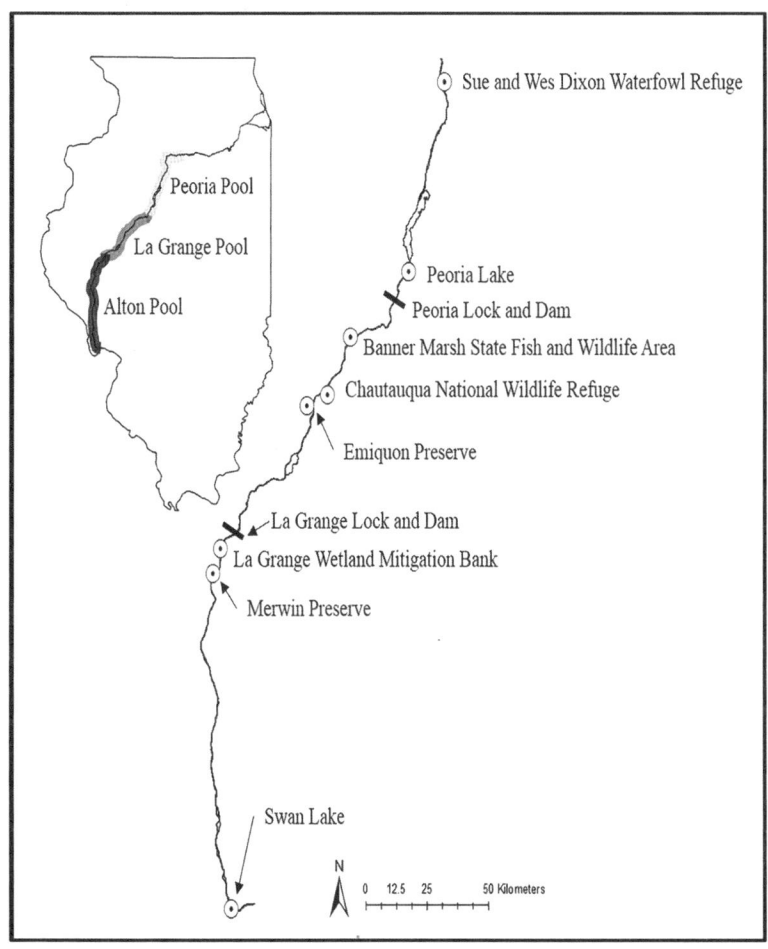

FIGURE 5. The middle reach of the Illinois River system, depicting floodplain wetland restoration and management sites.

IMPORTANCE, IMPACT, AND NEED FOR FLOODPLAIN WETLAND RESTORATION

When water is drained from or permanently added to a floodplain wetland, a multifunctional, highly productive, and unique natural environment is lost. In river systems important for commercial navigation (e.g., the Ohio, Illinois, and Mississippi Rivers), floodplains are often inundated permanently with surface water and not allowed to dry periodically. Despite the recent surge of interest in wetland restoration, significant gaps still exist in the application of theory to the restoration of these ecosystems, especially the link between the physical structure and ecosystem function. Restoration strategies often focus on re-creating historical forms with the intent of reestablishing desired ecosystem functions. However, accumulation of human impact on a landscape scale spanning centuries of time in large river systems makes achievement of historical function unlikely (Jackson and Pringle 2010).

In the early twentieth century, the Illinois River was modified to connect the Great Lakes to the Mississippi River and to facilitate commercial navigation (Delong 2005). Approximately half of the floodplain wetlands along the Illinois River were drained for agriculture by the 1930s (Middleton 2002), and encroachment, modified flooding regimes, and pollution from increasing urbanization degraded the river and its remaining floodplain wetlands (Bellrose et al. 1983; Sparks 1995). Interest has developed over the last half century to manage the Illinois River and its associated floodplain wetlands to support lost ecosystem services (e.g., flood abatement, nutrient processing, carbon sequestration, wildlife habitat), even if a return to the original river form and hydroperiod is unlikely.

There have been several notable floodplain wetland restoration efforts undertaken over the last forty years. Maturation and assessment of the diverse, large-scale restoration and enhancement projects along the Illinois River provide a unique perspective on best practices for floodplain wetland restoration projects and yield insight into the relationships between land management goals, public and stakeholder opinions, and the determination of a successful ecosystem restoration (Theiling 1995). Scientists are beginning to translate those restoration efforts into scientific theory, and managers are starting to incorporate an adaptive management approach to restoration while adopting practices for achieving targeted ecosystem services (Bajer et al. 2009; Sparks et al. 2017).

TABLE 1. Lessons learned and their implications for other future floodplain wetland restoration projects.

Hydrologic Connection	Lessons Learned	Restoration Implications
Open Peoria Lake	Uncontrolled, variable hydrology limits management capability and destroys infrastructure. Submersed aquatic and emergent vegetation failed to establish. Fish and waterfowl responded well to island building.	Rules of assembly and stability are influenced by presence/absence of a full suite of key structural habitats.
Partial Chautauqua National Wildlife Refuge LaGrange Wetland Mitigation Bank Swan Lake	Occurrence of floods in the wrong season degrades habitat and biotic response. Disconnection among potentially conflicting stakeholder goals can lead to failure to meet either perspective's needs. There is a minimum, and potentially nonlinear, threshold of restoration activity required to produce a measurable ecological improvement. A range of flawed engineering can lead to inadequate water management that, in turn, undermines the project in the long term.	An effective level of water control is needed when there is open but altered hydrology. Establishing the balance between wildlife and human needs should be done early in the planning process and should guide project activities. Project plans should be developed based on reliable information about the site. Plan for changing conditions in the future.

continued on next page

Hydrologic Connection	Lessons Learned	Restoration Implications
Limited Dixon Waterfowl Refuge at Hennepin and Hopper Lakes Banner Marsh State Fish and Wildlife Area Merwin Preserve Emiquon Preserve	Restoration planning and project goals need to consider the shifting baseline of species, especially the introduction of invasive species or ecosystem engineers. Water-level management and control can be just as important as in projects with open or partial connections. Non-focal species are important in developing a complete and healthy community and supporting the establishment of focal species.	Nuisance and invasive species can substantially alter restoration outcomes and require that project funds and resources be allocated for their control. The successful establishment of a desired species may be contingent on the dynamics of species outside the project site.

BUILDING ON THE LEGACY OF EARLY RESTORATION AND CONSERVATION EFFORTS

River-Floodplain Connectivity

The importance of the river connection to the floodplain cannot be over-emphasized as a driver of ecological form and function. However, this connection must be carefully considered from an economic and natural resource management perspective when developing restoration strategies for floodplain wetlands. It is crucial that trade-offs in ecosystem services from floodplain-river connectivity be understood so that the success, limits, and extent of restoration and conservation efforts can be appreciated. One of the main trade-offs is the balance between the need to protect biological diversity within floodplain wetlands from nutrient and sediment pollution carried by the Illinois River versus the need to establish floodplain connectivity for flood storage, nutrient and sediment removal, and other ecosystem functions.

River-floodplain hydrologic connections may be characterized as open, partial, and limited (Table 1). Open connections are those where surface water moves between the river and floodplain wetlands at normal river elevations or low-flow stages. Partial river connections occur at river levels intermediate between open and limited connections. For example, floodplain wetlands that are hydrologically isolated during low-flow and low-elevation floods but receive river water during moderate or severe floods are characterized as partially connected. Limited connections typically characterize floodplains isolated behind large levees (i.e., within drainage and levees districts) where water exchange does not occur between floodplains and the main river channel except through subsurface connections or during infrequent major floods (i.e., 100-year flood events).

The extent and duration of flooding shape vegetation communities in large river ecosystems. Open river-floodplain connections allow water exchange throughout the year, but they offer no protection from floods during the growing season and are often devoid of aquatic vegetation (Bellrose et al. 1983; Stafford et al. 2010). Partial river connections protect vegetation from low-magnitude flooding during the growing season, but they allow moderate- and high-magnitude floods to overtop levees and enter the floodplain. Partial river connections may maximize the diversity of ecosystem services provided by floodplain wetlands, but substantial variation in hydrology and resulting plant communities occurs across years and there will be trade-offs in the ecosystem services provided. When river levels are high during summer, flood storage and sediment trapping will be increased and extensive exchange of nutrients and other material will occur between the main river channel and the floodplain (Sparks 1995). However, prolonged high water during the summer can preclude growth of many emergent and submersed aquatic plants and substantially reduce wildlife habitat quality (Bellrose et al. 1983; Stafford et al. 2010; M. J. Lemke et al. 2017). In contrast, limited river-floodplain connections do not allow the historical flood pulse to reach floodplains (Junk et al. 1989), but hydrologic variability and major floods during the summer growing season are prevented, which allows establishment of emergent and submersed aquatic plants that provide high-quality habitat for wetland-dependent wildlife (Hagy et al. 2017; Hine et al. 2017).

Historically, spring and early summer floods from the Illinois River inundated floodplains, creating habitat for river fish and initiating a variety of important nutrient processes. During summer in most years, water receded from the floodplain, exposing mudflats and producing moist-soil vegetation. These natural drawdowns were an extremely important component of the wetland cycle and created important habitat for shorebirds, waterfowl, and other wetland-dependent wildlife. Bottomland hardwood forests composed of diverse mast-producing trees, such as oaks (*Quercus* spp.), and species adapted to wet-dry cycles of the river flood pulse were abundant and interspersed with areas of moist-soil vegetation and other aquatic plants. However, increased and stabilized water levels following installation of the lock and dam system resulted in loss of many mast-producing bottomland tree species and floodplain wetlands that filled with sediment.

A wide range of restoration projects with different goals have been implemented along the middle reach (e.g., Alton, LaGrange, and Peoria river navigation pools) of the Illinois River. Some restoration projects sought to restore structure and function of floodplain wetlands while maintaining extensive hydrologic connectivity to the main river channel, while other efforts focused on reestablishing a limited number of functions considered important for wildlife or fish using limited connections. The diversity of restoration projects provides an opportunity to extend ecological theory, especially within highly altered floodplain wetlands where predegradation conditions may not be known or attainable (Table 1).

Floodplain Wetland Restoration with Open Hydrologic Connections

Between 1903 and 2000, Peoria Lake lost more than two meters of mean depth and 68 percent of its volume to accumulating sediments from increased water levels, reduced flow rates, and increased sediment loads in the Illinois River (Bellrose et al. 1983; USACE 2001). The Peoria Lake Habitat Rehabilitation and Enhancement Program (HREP) through the U.S. Army Corps of Engineers (USACE) consisted of a four-phase restoration and enhancement project that added barrier islands along the river ridge and installed impoundments to reduce hydrologic connectivity and enhance water management capabilities. Although aquatic vegetation was planted on the barrier islands, vegetation establishment was unsuccessful. Stafford et al. (2010) demonstrated dramatic changes to vegetation communities along the Illinois River during the twentieth century and attributed lost submersed and floating-leaved aquatic vegetation communities to sedimentation, invasive carp, and fluctuating hydrology during the growing season. Waterfowl appear to have increased following barrier island construction, despite failure of vegetation establishment. Variable water levels of the Illinois River and levee heights that can be overtopped during the growing season hampered water management capabilities due to deleterious effects on infrastructure, such as levee erosion (USACE 2001).

Floodplain Wetland Restorations with Partial Hydrologic Connections

Chautauqua National Wildlife Refuge (CNWR) (17.8 square kilometers) includes a portion of a floodplain isolated from the Illinois River behind levees and managed for fish and wildlife habitat. From 1991 to 1998, the U.S. Fish and Wildlife Service partnered with the USACE through HREP to improve the water control infrastructure on this former drainage and levee district. They raised 6.1 kilometers of existing levees to heights that should exclude water one out of ten years during the growing season in a portion of the floodplain (i.e., the north unit) and one out of two years in another portion of the floodplain (i.e., the south unit) (USACE 1991). However, water levels and maximum stage heights in the Illinois River changed after completion of the project, and floodwaters typically enter the south unit every year and the north unit during most years for long durations during the growing season.

Although planners recognized that pumping to dewater the south unit would be necessary in most years, the feasibility of using expensive electric pumps was probably overestimated and costs created a barrier to completing timely drawdowns of the south unit. Additionally, invasive Asian carp entering the units through water control structures during dewatering created a large fish kill each year that caused negative public relations. Overall, efforts at the CNWR to create fish and wildlife habitat have been somewhat successful, but the ability to manage the dynamic flooding of the Illinois River was probably overestimated, despite an extensive planning process. While the improvements to develop infrastructure through HREP were significant, it failed to produce the desired control over growing-season flooding, timely drawdowns for moist-soil management in the south unit, and an isolated fishery with submersed aquatic vegetation in the north unit.

The LaGrange Wetland Mitigation Bank is a 6.7-square-kilometer former floodplain located at the confluence of the Illinois and La Moine Rivers.

The area was drained for agriculture in the early 1900s, isolated behind tall levees, and farmed until restoration began in 2001 (Brooks 2005). The Illinois Department of Transportation purchased this parcel to mitigate loss of wildlife habitat and floodwater storage, and restored approximately two square kilometers of forested wetlands and 3.2 square kilometers of emergent and scrub-shrub wetlands behind the original levees. Openings along the La Moine and Illinois Rivers allow floodwater to enter and exit the floodplain during high-water events. Although herbaceous annual moist-soil plants typically develop during summer drawdowns, passive management of water at this site usually failed to inundate these plants during periods of autumn migration of waterbirds.

Lack of water management capability limited the value of this restoration effort for fish and wildlife, and greatly reduced establishment of valuable bottomland trees. Yet the partial river connection allows typically higher spring river stages to flood vegetation established the previous year and promotes nutrient cycling, sediment trapping, and development of nursery habitat for young fish. Aerial survey data from the Illinois Natural History Survey indicated that spring flooding and rains make the LaGrange Wetland Bank an important foraging location for waterfowl and other waterbirds during migration (Illinois Natural History Survey, unpublished data). Additionally, during most springs and early summers, the Illinois River repeatedly loads dissolved nitrogen onto the floodplain through the partial river connections (Lemke and Jenkins 2006; Lemke et al. 2007). Nitrogen and other types of nutrient processing are often undervalued by policy makers and the general public.

Swan Lake (SL), in Calhoun County, Illinois, is one of the largest HREP projects on the Illinois River (11.7 square kilometers) (Garvey et al. 2007). The goals of this project were to restore aquatic macrophyte beds for migratory waterfowl and provide habitat for fish during winter and various spawning periods. Levees were erected along the river ridge and across the floodplain to divide the floodplain into three management units isolated from the river during moderate floods. However, during major floods, constructed spillways enabled floodwaters to enter the floodplain, creating a partial river-floodplain connection. Islands were constructed to help manage wind fetch, erosion control basins were installed to capture sediment, and water control structures were installed in the levees to allow native fish to access the floodplain wetlands (Garvey et al. 2007).

The principal lessons learned at SL were the necessities of critically evaluating the design of levees and river connection, and overplanning for vagaries of river flooding to ensure that historical wet-dry cycles occur. The initial water control structures and pump system were undersized and unable to remove water from the floodplain quickly, which prevented the sediment drying and compaction needed to benefit aquatic vegetation. Paddlefish (*Polyodon spathula*) and other important game fish (e.g., large-mouth bass [*Micropterus salmoides*]) failed to remain inside the connected portion of the backwater lake following release and did not migrate through the water control structure, although non-game fishes were common following restoration. Unprotected islands failed to reduce wind fetch and led to wave-driven suspension of sediment and extensive bank erosion that increased turbidity and degraded conditions for establishment of submersed and emergent aquatic vegetation over time (Garvey et al. 2007). Deep water areas typically relied upon by fish for seasonal habitat filled with sediment and fish failed to use the floodplain wetland during winter. This restoration effort reinforced the need for tall river levees to protect areas from moderate floods during the summer growing season where high quality wildlife and fish habitat are targeted.

Floodplain Wetland Restorations with Limited Hydrologic Connections

In 2001, the 12.1-square-kilometer Sue and Wes Dixon Waterfowl Refuge at Hennepin and Hopper Lakes was acquired by The Wetlands Initiative and restoration ensued behind an existing drainage district levee (Bajer et al. 2009). Restoration goals for the refuge included reestablishing pre-European wetland and associated upland vegetation communities. Prior to restoring hydrology, rotenone was applied to existing agricultural drainage ditches to euthanize common carp (*Cyprinus carpio*). Within six months after this treatment, precipitation, runoff, and seepages filled the historical floodplain wetlands and created an area exceeding five square kilometers dominated by submersed, floating-leaf, and emergent aquatic vegetation. Waterfowl quickly colonized the productive marshes, and more than 150,000 birds were observed during autumn migration (Bajer et al. 2009). However, common carp survived the initial rotenone treatments, and waterfowl and coverage of aquatic vegetation began to decline five years post-restoration and were mostly eliminated by seven years post-restoration.

In 2009, biologists once again pumped the floodplain wetlands dry and applied rotenone in ditches and low-lying areas to euthanize carp. Unfortunately, common carp used the existing subsurface tile drains as refugia from the rotenone treatments and aquatic vegetation communities again exhibited significant declines. Another complete drawdown was made from summer 2012 to spring 2013, and all subsurface tile drains were disabled before the site was treated with rotenone. By summer 2013, water was returned to the wetlands, more than thirty-four aquatic vegetation species colonized the newly wetted areas, and more than 500,000 waterfowl used the complex during autumn migration.

Despite extensive planning prior to restoration, the survival of common carp in agricultural drainage ditches and subsurface tile drains led to loss of a public fishery from 2009 to 2015 and millions of potential waterfowl use days from 2006 to 2013. In former agricultural areas, infrastructure to manage surface and subsurface water should be carefully designed before water is returned to sites. Rapid wetland restoration success can create false expectations or entrenched stakeholder groups that may oppose future changes to management or infrastructure (Sparks et al. 2017).

Banner Marsh State Fish and Wildlife Area (BMSFWA; 17.7 square kilometers) lies in a former floodplain of the Illinois River in Fulton and Peoria Counties, Illinois, and is isolated behind a former drainage district levee constructed in 1917 (USACE 2004). The drained area was used for farming and strip mining for coal prior to purchase and restoration by the Illinois Department of Natural Resources in 1978 and was converted to a complex of shallow marshes and deep-water areas that were formerly strip mines (Rickey and Anderson 2004). The lack of hydrologic connectivity between the created wetlands and the Illinois River limited nutrient exchange, flood storage, and nursery habitat for river fish. However, disconnection from the Illinois River allowed submersed aquatic, floating-leaf aquatic, and emergent vegetation to grow and provide valuable fish and waterfowl habitat. Invasive species (e.g., *Phragmites* spp.) have invaded many wetlands and may limit value of emergent cover for nesting waterbirds (Schummer et al. 2012). The former mines are deep, and drawdowns do not commonly occur, which limits productivity, but depth and the hydrologic isolation of these created wetlands provides a recreational fishery. BMSFWA illustrates the capability of developing emergent aquatic vegetation and game fish communities without a river-floodplain hydrologic connection. However, this site also illustrates limitations of wetland restoration in highly modified areas (e.g., strip mines) and in a floodplain that may have only provided limited habitat prior to levee installation relative to other floodplain wetlands (Havera et al. 2003).

The Merwin Preserve (4.7 square kilometers) in Brown County, Illinois, is owned and managed by The Nature Conservancy (TNC). This former Illinois River floodplain was leveed and drained in the early 1930s and farmed until 1998. Drainage infrastructure was dismantled at Merwin Preserve in 1999, and water was allowed to naturally accumulate behind the river levees, creating wetlands and associated habitats for wildlife and fish, and rehydrating the historic basin of Long Lake. Following cessation of pumping, extensive areas of submersed, floating-leaf, and emergent plants developed and waterfowl and sport fish flourished within this disconnected floodplain wetland. However, the abundance of submersed aquatic vegetation declined during the late 2000s as the stages of wetland succession advanced to the lake marsh stage (van der Valk and Davis 1978). A severe drought in 2012 resulted in a complete drawdown of Merwin Preserve, which produced extensive moist-soil vegetation and likely served to reset the marsh succession cycle.

In April 2013, floodwater overtopped levees and resulted in two breaches that enabled river-floodplain connectivity during high river water levels. The partial river connection resulted in conditions where spring and summer floods inundated the site for long periods during the growing season and river fish entered and exited through the river breach (Solomon et al. 2014). Although partial river connections may benefit shorebirds by providing mudflat habitat midsummer, minimal habitat exists for autumn-migrating waterbirds because of passive water level management. However, spring floods inundated the floodplain wetlands and provided important spring-migration habitat for waterfowl in one of every two years (M. J. Lemke et al. 2017). The partial river connections promote favorable conditions for denitrification, sediment removal, and floodwater retention, but they lead to decreases in the quality of wildlife and sport fish habitat.

A number of important lessons can be gleaned from Merwin Preserve. Attention to hydrology remains important in restoration efforts within river floodplains, even those that are not connected to the river. Managers must anticipate costs of moving water in and out of restored areas and install infrastructure to facilitate this movement. Also, it was noted in several sites that the recovery rate of aquatic vegetation communities compared to other ecosystems was relatively rapid and proceeded without supplemental planting, which will reduce up-front costs. Lastly, limited hydrologic connections undoubtedly produce the greatest quality fish and wildlife habitat, but the costs of levee maintenance, pumping, water control structures, and invasive species management are significant and must be anticipated up front (Sparks et al. 2017).

THE NEXT GENERATION OF ILLINOIS RIVER FLOODPLAIN WETLAND RESTORATION: THE EMIQUON PRESERVE

Creating Emiquon

Emiquon Preserve is a former drainage and levee district under agricultural production that, prior to levee construction and conversion, was a wetland-lake complex that was regionally famous for its abundant fish and wildlife populations (Havera et al. 2003). In 2000, TNC purchased most of the Thompson Lake drainage and levee district and discontinued drainage operations in 2007, thus allowing water to accumulate behind the river levees. The primary objective of the Emiquon restoration project was to restore natural ecosystem processes and habitats that promote and sustain the aquatic and terrestrial communities once found in the middle reach of the Illinois River. To achieve this objective, intensive planning began with the formation of the Emiquon Science Advisory Council and development of a suite of seventeen key ecological attributes (KEA) and sixty-seven indicators for different plant and animal taxa that could be used to gauge a restoration trajectory and determine when management might be needed (A. M. Lemke et al. 2017). Once a monitoring framework had been established, data collected, and indicators assessed, an ecological report card was used to communicate the condition of attributes and to determine the management actions needed (Harwell et al. 1999; Hagy et al. 2017). Establishment of a monitoring framework with key attributes (Figure 6), combined with systematically collecting data and summarizing results, increases the ease of informing stakeholders and the public of the adaptive management

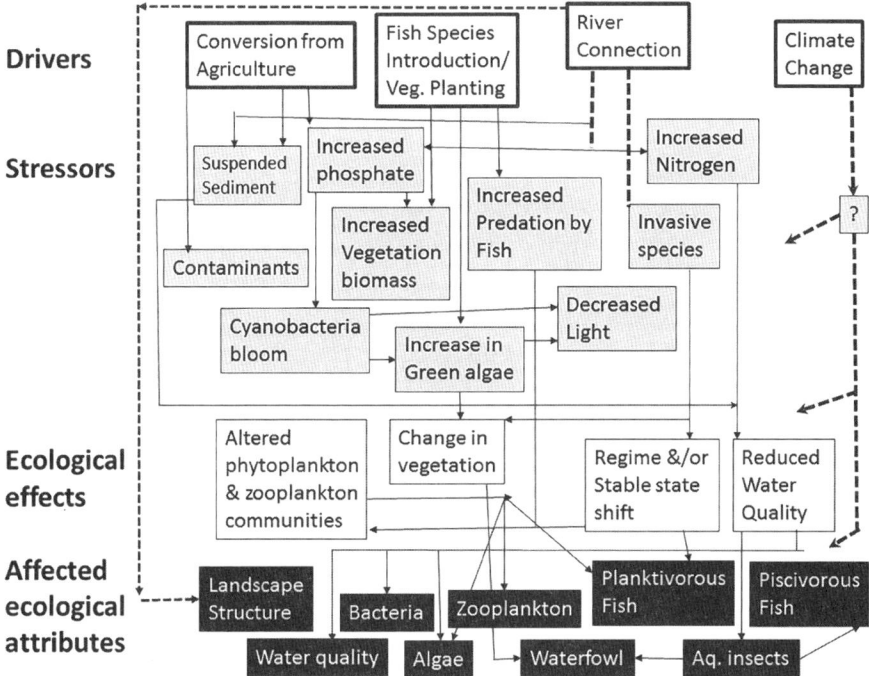

Drivers

Stressors

Ecological effects

Affected ecological attributes

FIGURE 6. A conceptual model showing links between ecosystem drivers, stressors, their ecological effects, and the affected attributes that can guide adaptive management at Emiquon Preserve. These attributes are part of a much larger list of key ecological attributes monitored by The Nature Conservancy and associated scientists at the Emiquon Preserve. The question mark denotes unknown impact. (Adapted from Galatowitsch 2012 and Harwell et al. 1999b.)

decision-making processes and is needed to evaluate high-profile floodplain wetland restoration efforts (Harwell et al. 1999; Sparks et al. 2017).

By 2009, the Thompson and Flag Lake areas of the Emiquon Preserve had grown from a network of agricultural drainage ditches in 2007 to a more than 19-square-kilometer biologically diverse wetland complex that is an important location for aquatic plants, native fish, and migratory birds (Hagy et al. 2017; Hine et al. 2017). A diverse mix of aquatic plant communities, including dense emergent, moist-soil, floating-leaved, and submersed aquatic vegetation, flourished in isolation from the unpredictable growing season floods of the Illinois River. Key to the success of the restoration effort was the joining of partners and stakeholders with common goals. TNC has developed partnerships with many state and federal agencies, universities, nongovernmental organizations, and over two dozen other entities to acquire, monitor, and manage Emiquon Preserve (Sparks et al. 2017). Subsequently, the Emiquon Preserve and the adjacent Emiquon National Wildlife Refuge were designated collectively as the Emiquon Complex and a Ramsar Wetlands of International Importance in 2012.

Since 2007, selected KEAs, as well as non-KEA organisms (e.g., algae, bacteria), were selected for monitoring due to their sensitive responses to multiple stressors and to help the Emiquon partners understand the ecological effects of restoration. These data and assessments provide invaluable baseline trends needed to assist the next phase of restoration, which will attempt to establish a controlled connection between the Emiquon Preserve and the Illinois River. An annual Emiquon Science Symposium facilitates the communication of monitoring and management updates among interested individuals and organizations involved with ongoing restoration efforts at the Emiquon Preserve. For more than a decade, close to 100 researchers, managers, and stakeholders have convened at this conference to publicly report on the status of their research, to evaluate ranges of indicators and attributes, and to address topics ranging from the history of the Illinois River to microorganisms that now teem in the waters of the Emiquon Preserve.

Lessons Learned from the Emiquon Preserve

Wetland and Aquatic Vegetation   During the initial phases of restoration at Emiquon Preserve, spatial coverage of aquatic vegetation and other aquatic cover types expanded from 2.5 square kilometers in 2007 to 19.4 square kilometers in 2013 (Hine et al. 2017). Diverse aquatic vegetation communities at Emiquon Preserve developed naturally during restoration with little supplemental planting or hydrological management (Figure 7). Vegetation emulated aquatic plant communities largely eliminated from floodplain wetlands in the Illinois and Upper Mississippi River valleys, with the most important among these types being the floating-leaved and submersed aquatic vegetation (Moore et al. 2010; Stafford et al. 2010). Although the vegetation response following return of hydrology was exceptional, maintaining consistent water levels over time can lead to reductions in aquatic vegetation coverage and diversity, overall primary productivity, and value

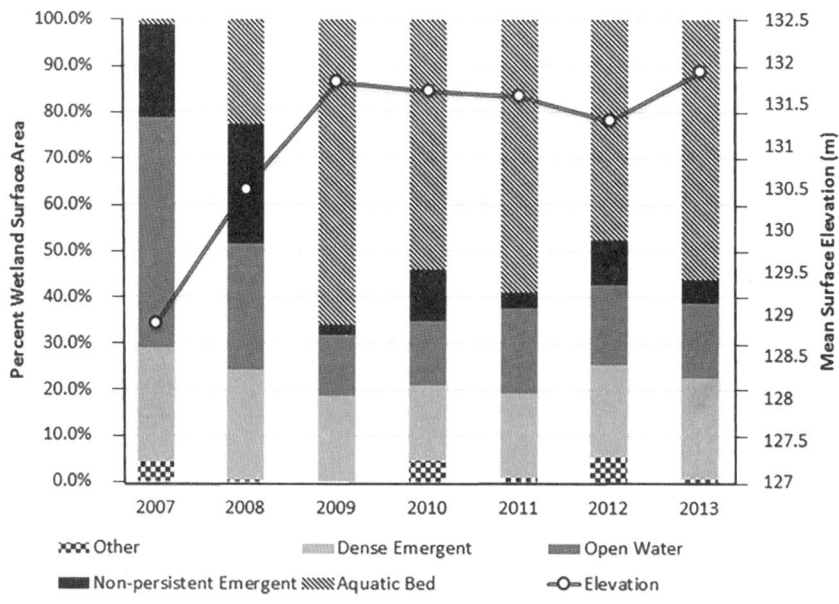

FIGURE 7. Percent coverage of aquatic vegetation communities and other cover types within the wetland area in relation to mean elevation of surface water at Emiquon Preserve from 2007 to 2013.

to wildlife and fish. In just six years, aquatic areas at the Emiquon Preserve entered the degenerating marsh phase typified by declines in emergent vegetation due to prolonged high and stable water levels (van der Valk and Davis 1978). Without a prolonged drawdown to expose and consolidate sediments and reverse the negative impacts on vegetation, habitat quality for wildlife and fish may decline substantially (Hine et al. 2017).

Biotic Community and Abiotic Condition Shifts   A clear-water state with high aquatic plant diversity is desired by many stakeholders in floodplain wetland restoration projects similar to the Emiquon Preserve because these conditions often produce greater diversity and productivity within higher trophic levels of plankton, waterbird, and fish communities (Hobbs et al. 2009). However, floodplain wetlands may undergo shifts between a clear-water state with an abundance and diversity of submersed and emergent aquatic plants to a turbid state where algae is dominant and aquatic

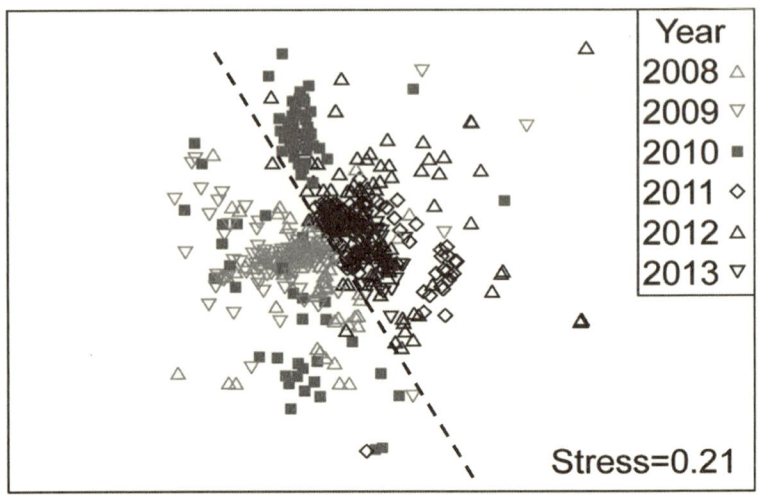

FIGURE 8. Nonmetric multidimensional scaling ordination of the pelagic bacterioplankton assemblages from Thompson Lake in the Emiquon Preserve. Each data symbol represents the assemblage composition (i.e., species number and relative abundance) for one sampling date, classified by year (see legend). The dotted line demarcates early restoration (2008–mid-2010; left of line) from later restoration (late summer 2010–2013; right of line). (Data source: M. Lemke, A. Kent, and S. Paver.)

plants are mostly eliminated (Scheffer 2004). State shifts are driven by three interconnected factors: (1) increased phytoplankton-derived turbidity due to nutrient increases; (2) aquatic plant competition with phytoplankton for nutrients; and (3) increased aquatic plant growth that decreases turbidity (Scheffer and van Nes 2007). The first phase of floodplain wetland recovery at the Emiquon Preserve was a response to an eighty-year disturbance (i.e., drainage and agriculture), and thus a mixture of abiotic and biotic factors drove the ecosystem. The recovering collective wetland area was unstable from 2007 to 2010 as the ecosystem structure and function were reestablished. A top-down control of the trophic cascade was amplified by the large initial stocking of a variety of piscivorous fish (Michaels and Sass 2009; VanMiddlesworth et al. 2017b) that allegedly would control invasive carp. Low levels of soluble nitrogen appeared to favor cyanobacteria blooms as early as 2008, though in subsequent years the cumulative aggregation,

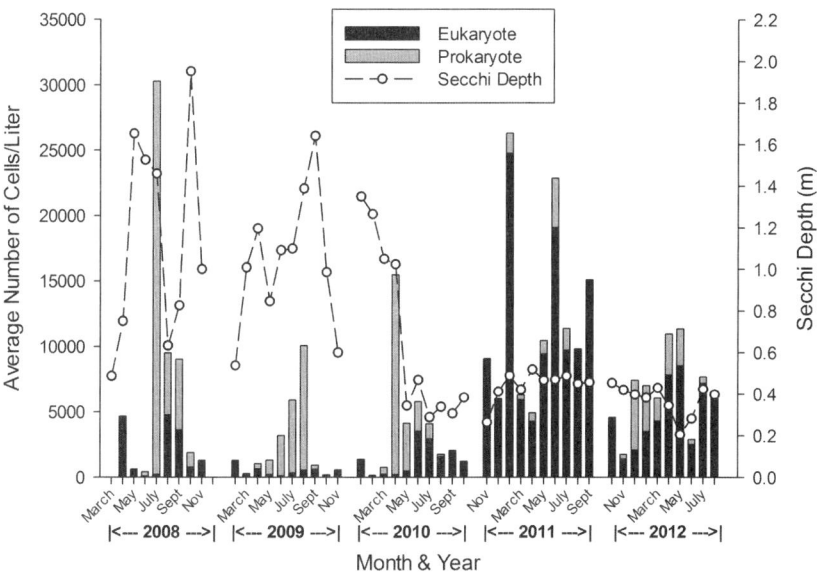

FIGURE 9. Water clarity measured by Secchi depth (right axis) and average abundance of prokaryotic autotrophic cyanobacteria and eukaryotic open-water autotrophic algae (left axis). (Data source: L. Rodriguez, M. Lemke, and K. Dungey.)

death, and decay likely contributed to a change in the nutrient dynamics of the three-year-old restored lake. Bacteria monitored biweekly revealed that bacteria changed in species composition and abundance in the summer of 2010 (Figure 8).

After 2010, the Emiquon Preserve wetland area exhibited a more turbid state and more stable conditions. Alternative stable ecosystem states occur when rapid changes from primary succession slow and productivity and biotic diversity plateau and eventually decline with increasing water turbidity (Jeppesen et al. 1990; Scheffer 2004; see Chapter 2). In Thompson Lake at Emiquon, the first years of restoration did not represent a stable state, but a period where the lake was in transition and adjusting to many changes. As prokaryotic autotrophic cyanobacteria lost dominance in the first two and a half years, eukaryotic open-water autotrophic algae became dominant (Figure 9). In this same time period, large-sized zooplankton decreased and gizzard shad (*Dorosoma cepedianum*) grew more readily, resulting in a more turbid wetland. Drivers known to cause a shift away from a clear-water state that supports diverse assemblages of aquatic plants include nutrient additions that create algal blooms, disruption of sediment by fish and waterfowl, extreme storm events, wave action, flood pulses carrying suspended solids, and drought (Bajer et al. 2009; Scheffer 2004).

Fish　As Thompson and Flag Lakes at the Emiquon Preserve expanded, shallow and deep water areas appeared along with diverse vegetation communities, resulting in a shift in the fish community. Initially, fish communities were dominated by the invasive common carp and goldfish (*Carassius auratus*) (VanMiddlesworth et al. 2017a). Early actions, such as rotenone treatment of ditches in 2007 and the stocking of large numbers of native piscivores (i.e., largemouth bass, bluegill [*Lepomis macrochirus*], black crappie [*Pomoxis nigromaculatus*]), reduced and subsequently limited undesirable species to 1 percent or less of the total catch (VanMiddlesworth et al. 2017b). In contrast, the fish species richness has grown consistently from seven species in 2007 to twenty by 2013 (VanMiddlesworth et al. 2017a). During this same period, the total biomass of native fish collected peaked in 2012 and remained well above initial levels. Abundance of nonnative fish has never been more than marginal. In contrast, the biomass of both native and nonnative fish has increased steadily through time, though nonnatives have never dominated. By 2013, gizzard shad were the most numerous species and large numbers of native piscivores comprised 40 percent of the total catch. Starhead topminnow (*Fundulus dispar*), a species of conservation concern in Illinois, has been consistently collected since stocking in 2008 (VanMiddlesworth et al. 2017a). These trends in fish community structure paint a clear picture of a recovering fish assemblage.

Waterbirds　During autumn and spring migration immediately following restoration, dabbling ducks and other waterbirds (e.g., American coots [*Fulica americana*]) showed the most dramatic responses. Each bird group accumulated more than three million use days per year and comprised more than 30 percent of the total waterbird use days in the Illinois River valley, despite the wetland survey area of Emiquon Preserve comprising only 1.0 to 6.7 percent of the total survey area from 2007 to 2013 (Hagy et al. 2017). Use of Emiquon Preserve by most herbivorous and granivorous species was considerably greater than the wetland area available in the rest of the floodplain wetlands surveyed by the Illinois Natural History Survey during fall migration of waterfowl through the Illinois River valley. Most striking was the immediate response of American coots and dabbling ducks such as blue-winged teal (*Anas discors*), green-winged teal (*Anas carolinensis*), northern pintail (*Anas acuta*), northern shoveler (*Anas clypeata*), and gadwall (*Anas strepera*). Several species of conservation concern (e.g., common gallinule [*Gallinula galeata*]) nested and raised young following restoration, likely due to rapid development of emergent aquatic plants and sustained water levels during summer resulting from hydrologic disconnection from the Illinois River (Hagy et al. 2017).

Carbon Cycle and Soil Organic Matter   Floodplain wetland loss results in a release of significant carbon from soil organic matter into the atmosphere (Euliss et al. 2006). Most of the flooded area within Emiquon Preserve is hydric soil (characteristic of wetlands) comprised primarily of silty clay, clay loam, and silty clay soils (NRCS 2017). Wetland restoration has the potential for carbon sequestration. After six years of restoration, the dominant aquatic plant community type at Emiquon consisted of aquatic bed, which included coontail (*Ceratophyllum demersum*). Additionally, the amount of cattails (*Typha* spp.) has steadily increased at Emiquon since 2007 (Hine et al. 2017). After six years of restoration, the total soil organic carbon (SOC) storage on the top forty centimeters was 48.03 milligrams per hectare (Mg/ha), while the SOC storage in a nearby cropland for the same depth was 45.02 Mg/ha (Buss 2014). The SOC storage of the restored wetland increased by 3.01 Mg/ha over six years at an average annual accumulation rate of 0.5 Mg/ha. Bernal and Mitsch (2012) found that the annual accumulation rate of SOC in temperate freshwater wetlands ranged from 1.12 Mg/ha in a mudflat riverine wetland to 4.73 Mg/ha in a forested wetland, which suggests wetland type can influence SOC accumulation.

SOC accumulation may correlate strongly with the development of many wetland attributes and the passage of time (Craft et al. 2003; Meyer et al. 2008). The restored wetland at Emiquon Preserve is at an early stage of restoration, and its total SOC storage is less than the 73 to 137 Mg/ha range exhibited by reference wetlands (Briddell 2012). With increasing restoration age, the SOC storage of Emiquon will likely continue to increase (Jones and Schmitz 2009; Buss 2014; Chen et al. 2017). Predictions based on the average annual accumulation rate of SOC suggest it may take about 50 to

178 years to accumulate SOC to the level similar to reference wetlands. This prediction is consistent with the conclusion of Ballantine and Schneider (2009), who estimated it will take a century for restored wetlands to attain carbon storage levels similar to a reference wetland.

Hydrologic Responses   In 2016, TNC installed a water control structure in the drainage and levee district levee to create a managed connection between the Illinois River and restored floodplain wetlands. A managed connection may enable TNC to maximize benefits of a limited and partial river connection by managing hydrology more precisely. By resetting wetland plant succession through the management of hydrology, wetland productivity and plant community composition will be cyclic due to a changing hydrologic regime (van der Valk and Davis 1978). Alternative wet and dry cycles can help manage invasive species, allow maintenance of levees and other infrastructure (e.g., drainage ditches, boat ramps, nesting platforms), and sustain high-quality fish and wildlife habitat in the long term (Bowyer et al. 2005; Bajer et al. 2009).

While average annual water level management is an important tool for managing floodplain wetlands, the seasonal flood pulse is just as important an ingredient for the establishment of a vital river ecosystem (Junk et al. 1989; Tockner et al. 2000). Each major flood in the United States tends to elicit more comprehensive flood management policies (i.e., floodways, floodplain storage areas) instead of structural solutions like raising levee heights. The inclusion of a more comprehensive set of management options for isolated floodplain wetlands would increase some ecosystem functions (e.g., removal of excess nutrients, flood storage) (Bernhardt et al. 2005) but simultaneously reduce others (e.g., wildlife habitat, agricultural production) (M. J. Lemke et al. 2017). For example, in 2011 the Birds Point–New Madrid Floodway was used as an upstream diversion to limit rapidly rising waters near the Birds Point area to protect downstream levee districts and Cairo, Illinois, from flooding on the Mississippi and Ohio Rivers. While agricultural areas were sacrificed, this action inadvertently benefited natural resources (Olson and Morton 2013; Phelps et al. 2015).

River flooding can influence restored floodplain wetlands differently (M. J. Lemke et al. 2017). Floodwaters overtopped existing levees at Emiquon Preserve during the largest flood on record in the area in 2013 (Figure 10) and inoculated the wetlands with organisms and nutrients. However, the

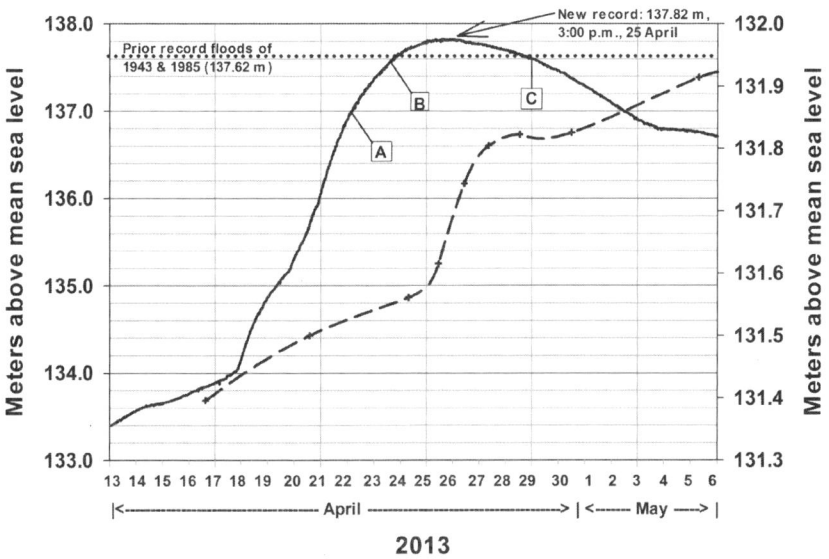

FIGURE 10. Water surface elevations of the Illinois River at Havana, Illinois (solid line; left y-axis) and Thompson Lake at the Emiquon Preserve (dashed line, right y-axis) for the 2013 Illinois River flood. Water enters Thompson Lake as North Globe levee overtopped at 137.0 meters (A); Water enters over South Thompson levee at 137.6 meters (B). River level recedes below levee top (C). (Adapted from Blodgett 2013; used with permission from The Nature Conservancy.)

existing levee remained intact and buffered the wetland against extreme water level change. In contrast, the same flood caused levees to breach at Merwin Preserve and allowed an open two-way connection between the Illinois River and its floodplain. While biotic and abiotic changes were minimal at Emiquon Preserve, the restored wetland was dramatically altered at Merwin Preserve, illustrating the trade-offs between wildlife habitat and vegetation diversity in isolated wetlands versus floodwater storage and nutrient exchange in connected floodplains.

The sudden flood that results behind a breached levee differs from the gradual rise of the typical seasonal flood and may set back restoration efforts, deposit silt, cause scouring that hampers crop production, introduce invasive species, prolong water residency, and require extensive infrastructure repairs (Tobin 1995). Initially, vegetation and microbial communities inside a recently connected floodplain wetland provide extensive ecosystem services, but these communities likely degrade quickly over time as aquatic plants are eliminated and influxes of sediment increase flocculence of the substrate. Thus, there can be short-term benefits in nutrient removal and even wildlife habitat following floodplain reconnection, but over time, these ecosystem functions will likely be much reduced with an open or partial river-floodplain connection (M. J. Lemke et al. 2017).

Flooding brings nutrients, invertebrates, seeds, and other materials into floodplain wetlands. Therefore, hydrologically connected floodplains can increase productivity of river ecosystems (Junk et al. 1989), but these benefits cannot be realized if frequent disturbance inhibits natural succession (Robertson et al. 2001; Jackson and Pringle 2010). Of considerable note is that unregulated connection to rivers with greatly altered water and sediment regimes can favor invasive species and degrade conditions for aquatic plants (Galat et al. 1998; Tockner and Stanford 2002). We hypothesize that a managed river-floodplain connection could maximize ecosystem functions and species diversity without degrading wetlands, but large-scale experiments are needed to verify this hypothesis.

CONCLUSIONS: INTEGRATING ECOLOGICAL THEORY INTO THE PRACTICE OF FLOODPLAIN WETLAND RESTORATION

Floodplain wetland restoration may not always proceed as planned or proceed to the projected restoration target, but more likely, restored floodplain wetlands move toward exhibiting a range of conditions regulated by the frequency of disturbance and presence of ecosystem drivers. We have illustrated different aspects of the restoration progress and compared floodplain wetland restoration projects from a variety of sites along the Illinois River. An emphasis on the importance of strong partnerships and long-term commitments in monitoring and assessing the outcome of restoration projects to advance the field of ecological restoration cannot be stressed enough. While restoration to guide recovery of ecosystem structure and function is the ideal, individual goals may be more achievable and could include

(1) compensation for lost historical ecosystem functions; (2) delivery of target ecosystem services under conditions present at the site or within the region; and (3) an endpoint that ensures resilience to perturbations, disturbances, and stressors, which are inevitable (Suding 2011).

Ecological restoration is a relatively new and growing discipline. The definition of ecological restoration is the science and practice devoted to assisting the recovery of an ecosystem that has been degraded, damaged, or destroyed (SER 2004). One of the most difficult phases of restoration is the planning phase. Goals and objectives must be clearly articulated, prioritized, and agreed upon by a diverse cadre of stakeholders. The limitations of infrastructure and resources for management must be clearly identified, and conditions within the restoration site and adjacent influences must be anticipated. The latter can be particularly challenging, as there is no singular process for anticipating changes in climate, surrounding land use, or other factors that might influence conditions within the site.

Often these goals are accomplished through a combination of linked management actions based on theoretical relationships between the potential actions and the desired state (e.g., degraded floodplain wetlands need inundation to begin succession toward greater diversity and productivity). However, there is a great deal of uncertainty inherent in applying theory in practice, especially related to myriad small variations in site-specific conditions (e.g., elevation, soil type, distance from nondegraded wetlands), contingency (e.g., type of land use prior to restoration, hydrologic and climate cycles), and restoration goals (e.g., ecosystem versus single species).

A recurring theme embedded in floodplain wetland restorations along the Illinois River was maintenance of a full or partial hydrologic connection to the river. Restoration plans and goals consistently overestimated the likelihood that aquatic vegetation could establish and persist and underestimated the deleterious effects of extreme hydrologic variation on infrastructure. Moreover, despite predictions (Bellrose et al. 1983) that continued sedimentation would increase the magnitude of flood heights and frequency of flooding, an extensive planning process with hydrologists and engineers was insufficient to provide conditions capable of producing target wildlife habitat goals in most years. Thus, one of the most important aspects of river floodplain restoration is developing goals and targets compatible with infrastructure capabilities and long-term predictions of ecosystem changes.

In some instances, trade-offs in wetland functions and values must be predicted a priori and a prioritized set of goals dictated so that changing hydrologic conditions and inevitable infrastructure failures can be incorporated into the process. For example, restoration projects with limited goals can produce a functional wetland that provides flood storage, nutrient processing, groundwater recharge, and other ecosystem functions (e.g., LaGrange Wetland Mitigation Bank). Trade-offs exist between maximizing these functions while balancing fiscal requirements to produce sustained and high-quality wildlife habitat, which can still be uncertain despite extensive up-front planning and investment (e.g., SL, CNWR).

Logically, floodplain wetlands with minimal infrastructure will suffer fewer infrastructure failures. However, floodplain wetlands with minimal infrastructure may also act as sediment traps and gradually fill over time. Lower levees permit greater inflow and outflow frequency, promote more nitrogen processing through diverse flood cycles, and will mitigate the sediment trap retention role of the wetland to prolong the functional lifetime of the wetland. However, increased frequency and duration of flooding in floodplain wetlands also likely limits growth and diversity of aquatic vegetation that is valuable to wildlife and fish. What may appear as an economical approach to river and floodplain management of the Illinois River and other highly modified rivers in the Midwest are minimal time and capital investment associated with low levees and high river-floodplain connectivity. However, this approach typically compromises waterfowl and sport fish habitat, as well as other recreational values for the general public.

A more parsimonious approach might be to provide a mix of disconnected and partially connected floodplain wetlands managed cooperatively at large spatial scales (e.g., watersheds, ecoregions). Cooperative projects that leave portions of floodplains disconnected from the Illinois River will provide other wetland functions and values, such as wildlife and fish habitat, while other projects with partial connection will allow floodwater storage, nutrient processing, and fish nursery habitat. Since all wetlands and floodplains cannot simultaneously provide all functions, working collaboratively within partnerships (e.g., the Emiquon Complex) to restore floodplain wetlands over large spatial scales will provide a great variety of functions and values without constant examination and evaluations of trade-offs within individual sites.

Managing floodplain wetlands in complexes may also ameliorate public perception issues when changes in management result in decreased public access or changing target conditions or functions over time (Sparks et al. 2017). Restoration of floodplain wetlands is beneficial to waterbirds, and changing conditions or staggered vegetation community succession stages should ensure habitat availability within a region for many different guilds of wildlife and fish. A large river restoration culture has emerged along the middle reach of the Illinois River that includes conservation entities (e.g., TNC, The Wetlands Initiative, USDA Natural Resources Conservation Service), members of the public, hunters and anglers, and academic institutions. Thus, managing wetland cooperatively between agencies and in complexes ensures that public access is always available in at least some portions of the complex and can contribute to teaching and outreach efforts.

Nearly a century ago, lands were being set aside from development in a time when resource conservation was meeting the science and application ideas of game management (Leopold 1933). For decades, the restoration of degraded ecosystems was undertaken mostly by natural resource management agencies focused on anthropogenic uses of the environment. The historical restoration efforts for the Illinois River follow this trend in that floodplain wetland restoration efforts at that time were closely related to human manipulation of resources (e.g., application of farming practices for conservation, forest harvesting, nutrient control, introduction of desirable plants and animals). A union of industry, transportation, and recreational users of large river systems is logical because of the common philosophy of encouraging multiple resource use among land management agencies. Recently, the field of ecological restoration has indicated a shift in its vision from purely ecocentric visions of restoration to those compatible with anthropogenic land uses (SER 2004).

With respect to ecological restoration, the Emiquon Complex has built a foundation that combines observational science, modern resource management practices, and operational aspects (i.e., planting, hydrologic manipulation). This in itself within a large-scale effort in a highly dynamic and not yet well-understood ecosystem is laudatory. What is significant to the science and theory of restoration ecology is what happens next—how lessons learned are applied, if monetary resources will follow management decisions, and if science can take the next steps to apply hypothesis testing to decades-old questions to obtain new insights to ecological restoration. The science of ecology and limnology gives us the paradigms of succession, alternative stable states, nutrient budgets, and trophic cascades to work with (see Chapter 2). It is doubtful that any one paradigm in its present state will describe degraded or partially degraded ecosystems undergoing restoration. An approach that would help to answer issues under debate would be to conduct experiments that would test specific hypotheses, as opposed to relying on observation-based science. Long-standing issues that often feed misguided interpretations would have evidence that would settle some of the issues and fuel science theory. The challenge, excitement, and potential of the river restoration culture on the Illinois River is to apply what has been done and observed at the center of these impressive conservation and restoration efforts to refine theories and, in the long run, improve our predictive power (Zedler 2000a; Hobbs and Harris 2001).

CHAPTER FIVE

RESTORING STREAM ECOSYSTEMS IN THE MIDWEST

LUTHER AADLAND, NEIL HAUGERUD, AND CHRISTIAN LENHART

INTRODUCTION

Ecological restoration of streams in the Midwest has been conducted primarily since the early 1990s, although attempts to manage and "improve" them for human benefit and fish habitat enhancements have a long history in the region. Since European colonization in the 1800s, most historical management efforts have focused on installing dams, channel straightening, dredging, armoring, and other physical habitat modifications of streams. Additional impairments to rivers have been caused by point and non-point water pollution, unsustainable land-use changes, flow regulation with dams, wetland drainage and the resulting hydrologic changes, and climate change. While some impacts to lotic ecosystems may only be corrected with broad and substantive changes in public policy and land stewardship, others can be remediated with focused river restoration projects.

HISTORICAL IMPACTS TO MIDWEST STREAM ECOSYSTEMS

Stream ecosystems are defined by components of hydrology, fluvial geomorphology, water quality, connectivity, and biota (Annear et al. 2004). Each of these components comprises groups of variables that interact with each other in complex ways. Therefore, alterations to any of the components can have cascading effects on streams. Hydrology in particular is often seen as the key driver in stream ecosystems, since flow shapes the stream's physical environment, which in turn shapes the biological environment (Bunn and Arthington 2002).

Many problematic past alterations were based on the pursuit of narrow or singular anthropogenic goals such as drainage, conveyance, water supply, or power production. The projected outcomes of these projects were often based on simple and localized hydraulic models that did not account for complex watershed-scale effects on hydrology, geomorphology and habitat, biotic migrations, water quality, or consideration of the impacts on aquatic biodiversity. The primary goals of most channel-related management in agricultural watersheds of the Midwest was, and still is, to enable cultivation of former wetlands, increase agricultural productivity within poorly drained soils, and to increase capacity of stream channels (Dahl and Allord 1997).

Mathematical and physical models such as the Hydrologic Engineering Center's River Analysis System (HEC-RAS) have become increasingly sophisticated but have primarily focused on hydrology and hydraulics and have not attained the complexity necessary to accurately predict ecosystem responses of rivers. The failure to assess complex geomorphic and biological responses and to verify responses empirically in situ have led to unanticipated consequences for river projects. For example, stream channelization was long referred to as "channel improvement" by engineers and farmers. Not until later were the devastating effects on habitat and biodiversity, upstream channel incision and erosion, downstream sedimentation, and increases in downstream peak flow fully recognized (Shankman and Pugh 1992; Wilcock and Wilcock 1995; Urban and Rhoads 2003; Schlosser 1987; Pilcher et al. 2004).

Past alteration of streams has significantly affected ecosystem functions. While drainage and channelization were often done for flood control, these projects collectively cause increases in runoff and flood flows downstream that in turn cause channel incision, increases in erosion, turbidity, and sedimentation (Schottler et al. 2013). The Wild Rice River in northwest Minnesota was straightened by the U.S. Army Corps of Engineers in 1955 and subsequently caused upstream downcutting of its bed by four meters, completely separating it from its floodplain and causing major erosion problems that resulted in 2.5 meters of aggradation at the downstream end of the project (ACOE 1988, 2004, 2006). Channelization and other traditional engineering practices are a primary reason why many streams are in need of restoration. In Minnesota, 25,860 kilometers (49.1 percent) of streams have been channelized, 1,844 kilometers (3.5 percent) have been impounded, and only 24,916 kilometers (47.4 percent) remain "natural" (Lundeen 2014). Estimates of the amount of channelized streams in Iowa are even greater; there, nearly 80 percent of the streams have been channelized (Arbuckle et al. 2004). Similarly, dams were built for flood control and recreation, but they block migrating fish and other aquatic animals; inundate critical habitat; cause upstream sedimentation and downstream incision; alter nutrient, temperature, and flow regimes; and can cause catastrophic floods when they fail.

Alteration of nutrient cycles and sediment loading rates have had profound impacts on stream ecosystems and overall water quality. Stream channelization, reductions of riparian habitat, loss of headwater and floodplain wetlands, and the installation of subsurface tile drains have led to increases in nutrient and sediment input into Midwest rivers. The biota of small streams are effective in processing nutrients, and both phosphorus and ammonium were typically taken up in less than twenty meters of channel length in unchannelized headwater streams (Mulholland et al. 2000). In contrast, nutrients are not assimilated efficiently in agricultural drainage ditches. Within drainage ditches, Ahiablame et al. (2011) measured ammonium uptake over lengths of up to 84,650 meters and phosphate uptake over lengths of up to 22,700 meters, with no significant uptake in most cases. Nutrient problems are exacerbated downstream and become increasingly difficult to manage locally because nutrient assimilation rates decrease as rivers become larger. Nitrogen assimilation has been found to be over 60 percent of watershed yield in small streams but less than 10 percent on large rivers (Alexander et al. 2000).

Addressing nutrient inputs and sediment loading has been a major goal of Clean Water Act–funded programs in many states in the Midwest. Watershed management projects to reduce nitrate, phosphorus, and sediment loads often target the improvement of conditions in larger downstream waterbodies such as the Mississippi River and Gulf of Mexico, and require intensive management of upstream watersheds to be successful (EPA 2015). Eutrophication is problematic in many rivers and lakes in the Midwest that receive inflow from nutrient-laden ditches and streams. For example, the Maumee River is a State Scenic River in northwest Ohio that drains into

Lake Erie. The Maumee River exhibits excessive nutrient loading from upland agricultural use that causes serious cyanobacteria blooms (Kane et al. 2014). Stream sedimentation has been considered to be one of the major causes of reduced biodiversity and habitat loss in streams (Waters 1995). Excess sediment fills the hyporheic zone with fine sediment and reduces the physical habitat for aquatic macroinvertebrates, small fish, and glide spawning fish. Sediment is delivered from surface runoff over agricultural fields and from in-channel processes such as bed downcutting due to downstream channelization. A channel that has downcut substantially is disconnected from the floodplain and initiates lateral erosion that delivers even more sediment downstream as a new floodplain is formed. Lateral connectivity is an important water quality attribute, as significant nutrient exchange and processing occurs in the floodplain (Junk et al. 1989).

Fragmentation and degradation of habitat in streams have led to changes in fisheries, biodiversity, and the prevalence of invasive species. Rockström et al. (2009) concluded that the loss of biodiversity was the most severe environmental decline facing humanity. Alarmingly, the extinction rate of freshwater organisms in North America (3.7 percent loss per decade) is estimated to be five times the rate of terrestrial organisms (0.8 percent loss per decade) (Ricciardi and Rasmussen 1999). Aquatic species loss has been substantial in the Midwest as well (Karr et al. 1985). Based on recent surveys, the Minnesota River watershed has likely lost twenty-one of its forty-three native freshwater mussel species (Sietman 2007), with dead shells as the only sign of their former existence.

Habitat loss and fragmentation have had major effects on biodiversity in the Midwest. Good-quality stream habitat may lack native fish and freshwater mussel (family Unionidae, family Sphaeriidae) species if downstream barriers prevent spawning and seasonal and post-drought migrations. Conversely, a stream free of barriers to fish migration may lack native species if the habitat is degraded by channelization or inundated by reservoirs.

HISTORY OF MIDWEST STREAM RESTORATION

Numerous "habitat improvement" projects targeting trout (subfamily Salmoninae) were built in the 1930s (Gee 1952) and continue to be a prevalent strategy. These early efforts focused on habitat structures and often failed to consider fluvial process, appropriate morphology for the channel, and underlying causes of degradation. In many cases, the goal of "habitat improvement" was not to restore the morphology of natural channels but to create fish habitat with use of instream structures (Gee 1952). The lack of geomorphic context for many of these projects and the eventual deterioration and failure of structures makes structure-dependent approaches reliant on ongoing maintenance and repair (Frissell and Nawa 1992). In addition, quantitative evaluation of fish community, geomorphic, and ecological effects of habitat structures were rare, as most previous studies focused only on trout (Thompson and Stull 2002).

Efforts to restore habitat and natural characteristics of degraded streams go back at least 135 years. Van Cleef (1885) described streams that had been damaged by overfishing and removal of riparian trees and shrubs. He outlined practices that could contribute to stream restoration that included (1) protecting riparian trees and other woody vegetation; (2) protecting pools having abundant woody vegetation; (3) planting trees on the riverbanks wherever feasible; (4) placing stumps and branches in the pools for cover; and (5) prohibiting fishing with bait.

Stream restoration based on the morphology of intact natural channels (i.e., reference ecosystem; see Chapter 2) began in North America by the late 1970s (Newbury and Gaboury 1993) and within Europe by the 1990s (BSMRDE 1997). Leopold et al. (1964) advanced the understanding of natural channels and fluvial process that provided a foundation for environmental planning (Dunne and Leopold 1978) and stream restoration. A classification system for natural channels (Rosgen 1994) and a systematic approach to river restoration known as natural channel design have been developed (Rosgen 1993; see Chapter 9). Stable natural reference reaches serve as physical models for river restoration within the natural channel design approach. Specifically, stability is defined as allowing the river to develop a stable dimension, pattern, and profile such that, over time, channel features are maintained and the stream system neither aggrades nor degrades (Rosgen 1996).

Since many fundamentally disparate activities have been categorized as stream restoration, a precise definition is important. Ecological restoration is differentiated from habitat improvement or habitat enhancement projects, as these projects typical use designs that are not based on natural channel geometry measurements and depend on built habitat structures that require maintenance and will ultimately deteriorate. In simple terms, restoration means to bring back to an original state, but streams are constantly changing, which makes the original state difficult to define. Stream restoration needs to not only address the importance of channel form but also to acknowledge the physical, chemical, and biological processes that make up a healthy stream. We define stream restoration as the act of relaxing human constraints on the development of natural patterns of diversity where restoration measures should not focus on directly re-creating natural structures or states but on identifying and reestablishing the conditions under which natural states create themselves (Ebersole et al. 1997; Frissell et al. 1997; Frissell and Ralph 1998). This is comparable to the Society for Ecological Restoration definition, which is "assisting the recovery of an ecosystem that has been degraded, damaged, or destroyed" (SER 2004). Successful stream restoration involves the following: (1) determination of underlying causes (usually anthropogenic) that led to degradation and the need for restoration; (2) reestablishing natural channel morphology appropriate for the geology, slope, sediment supply, hydrology, valley type, climate, and ecology of the reach; (3) reestablishing fluvial processes that will continue to develop habitat in the channel and maintain floodplain connectivity; and (4) reestablishing biodiversity comprised of living organisms that stabilize the channel (plants and freshwater mussels), process nutrients, habitat, and food for other terrestrial and aquatic organisms.

The use of natural channel morphology as a physical model is an inherent element of stream restoration. Simply excavating a channel similar to a natural channel and locking it in place with structures does not restore critical ecosystem processes that depend on the ability of a stream to adjust and rebuild floodplains and habitat. By reestablishing fluvial and ecosystem processes, ecological restoration is self-sustaining. It is noteworthy that pristine streams rarely require restoration, because fluvial and ecosystem processes are already intact. This does not imply that streams are or should be designed as static ecosystems. In fact, processes of channel migration, floodplain formation and maintenance, resilience to climatic variability and change, and dynamic interactions with biological processes are key ecosystem functions that reflect the dynamic nature of lotic ecosystems.

If properly applied, a finished stream restoration should be difficult to distinguish from a natural channel. Unfortunately, some projects have overused structures to the point where they bear little resemblance to a natural channel.

Man-made synthetic materials, such as geotextile fabric, steel cable, steel mesh, and duck-billed anchors, will ultimately be exposed, are unaesthetic, and can cause a number of problems. Geotextile fabric, which is widely used in river projects in the Midwest, has been noted to (1) prevent root penetration by riparian vegetation and create failure planes that can result in stream bank failure; (2) block movement of benthic aquatic macroinvertebrates and the flow of oxygenated water through the streambed when used in riffles and log structures; and (3) lack the self-healing properties of granular filter material (FEMA 2006). By limiting materials used in restoration to native wood, stones, and vegetation, and using designs that mimic natural features, these structures are more likely to look and function as they do in unaltered natural channels. Structures will continue to be a necessary part of stream restoration for initial bed and bank protection of newly constructed channels and for protection of structures in urban projects, but their use should be limited and temporary. Ultimately, structures should be supplanted by subsequent recruitment of bedload, woody debris, and riparian and submergent vegetation to provide grade control and bank protection.

The goal of stream restoration should not be to create a static, immoveable channel but to restore dynamic natural processes that include channel migration and associated floodplain formation and maintenance. These processes also allow for adjustments resulting from changes in climate and watersheds. For instance, many forested streams in the Midwest possess terraces that formed following the clear-cuts in the late 1800s and the subsequent channel incision (Verry 2004; Anderson et al. 2006). While this disturbance caused significant instability and sediment yield that can be identified in lake sediments, many of these streams adjusted to a new, lower grade and established new floodplains. This resilience is critical to long-term stability and ecosystem function.

Evaluation of stream restoration projects is a critical step in determining ecosystem benefits and guiding future project design in terms of geomorphic stability, connectivity, hydraulics and hydrology, water quality, biodiversity, and other ecological responses. Replicates and controls incorporated into a before-and-after-control-impact study design are fundamental to assessment of effects of stream and river restoration projects (Stewart-Oaten et al. 1986; Smith et al. 1991).

Our objective is to discuss restoration of natural channel morphology in channelized rivers and restoration via dam removal and the associated problems. These two areas of stream restoration are emphasized because channelization and dam construction are the most profound causes of stream impairment that often require restoration. Technical details of the design process are beyond what can be fully covered in this chapter, but they will be briefly described in the case studies. The following case studies will provide (1) historical context of the sites and alterations that lead to the need for restoration; (2) a summary of the design challenges and approach; (3) a summary of construction issues; and (4) a summary of assessments and ecosystem responses. The Lawndale Creek restoration and the Appleton dam removal are projects implemented by the Minnesota Department of Natural Resources and designed by the lead author that are representative of restoration approaches used for many other similarly impaired rivers in Minnesota. Additionally, an assessment of five dam removal projects in Wisconsin is discussed to characterize post-dam removal sediment and vegetation conditions, and identify related management issues.

RESTORATION OF A CHANNELIZED PRAIRIE STREAM: LAWNDALE CREEK

Lawndale Creek is a spring-fed prairie stream in west central Minnesota that flows over Campbell Beach of glacial Lake Agassiz. The stream was channelized over much of its length starting in the 1890s. The project site is located within the Atherton State Wildlife Management Area in Wilkin County where two ditches (State Ditch 14, County Ditch 40) replaced the original stream and drained wetlands (Figure 11). The catchment has a drainage area of thirty-six square kilometers at the upstream end of the management area and forty-four square kilometers at the downstream end of the reach. State Ditch 14 was channelized in the 1890s, and County Ditch 40 was channelized in 1960. County Ditch 40 carried all base flow and most flood flows within a trapezoidal channel that lacked riffles, pools, or a connected floodplain. The straight County Ditch 40 had a steeper slope than the sinuous channel it replaced, resulting in greater erosive force. This caused 1.2 meters of incision and perched the upstream culvert, creating a partial barrier to fish passage. Lawndale Creek was listed by the Minnesota Pollution Control Agency as turbidity impaired, which is an indicator of high sediment or organic concentrations. The fish community in County

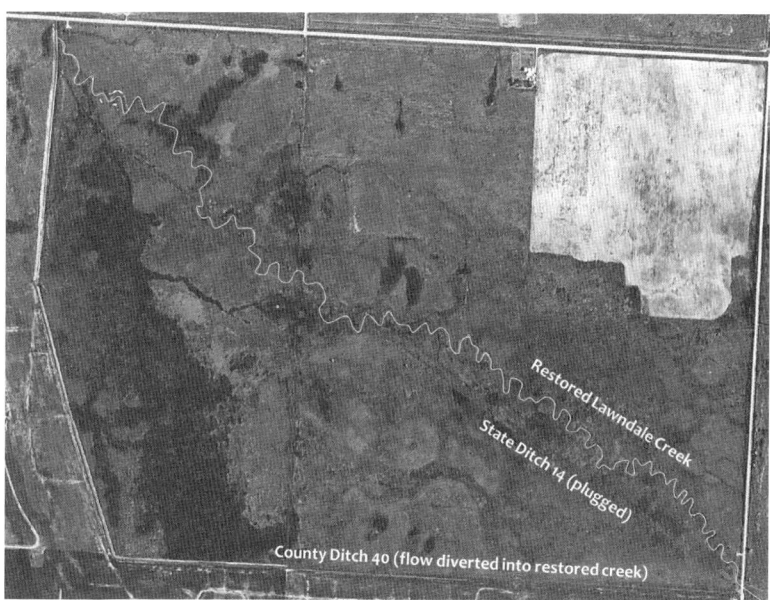

FIGURE 11. Aerial photo of the Atherton Wildlife Management Area showing the restored Lawndale Creek channel and riparian wetlands.

Ditch 40 was dominated by small-bodied minnow species (family Cyprinidae) associated with the homogenous shallow runs and lacked piscivores and larger pool-oriented species, typical of streams with more variety of depth and hydraulic conditions.

Interest in restoring the stream led to data collection needed for design and quantitative assessments of the fish community. Two natural reaches of the stream served as reference sites for channel geometry, hydraulics, and habitat. The upper reference reach was adjacent to native prairie in the Rothsay Wildlife Management Area and was narrow, deep, sinuous, and stable with a connected floodplain, diverse habitat, and a channel slope of 0.3 percent. The lower reference reach was near the confluence of Lawndale and Deerhorn Creeks on private land, and it was sinuous and relatively stable with a connected floodplain and a slope of 0.1 percent. Despite the high-quality habitat characteristics, the lower reference reach was affected by agricultural runoff. These sites were surveyed for channel geometry at the bankfull stage. The bankfull stage was determined primarily on the flat surface of the point bars, where sediment deposition is actively building the floodplain as channels migrate over time. Channel cross sections of riffles and pools, and reach profiles including thalweg, water surface, bankfull stage, and low bank height (which indicates floodplain connectivity) were surveyed in the field, while meander pattern geometry (sinuosity, radii of curvature, meander belt width, etc.) was measured from aerial photographs.

Sediment composition of the stream was evaluated using several methods. Pebble counts were taken to determine streambed composition, and point bars were evaluated to determine composition of transported sediment and assure the channel design could handle the sediment load. Bar samples were later verified by bedload and suspended sediment measurements. Fine sand dominated the streambed and bedload, with silt and very fine sand dominating the suspended load. Fine gravel represented the largest particles transported at bankfull discharge in the restoration reach.

Discharge measurements were taken in the reference channels to empirically determine bankfull flows and associated hydraulic coefficients. Field verification of channel hydraulics was important for designing, modeling, and permitting of the project. A hydraulic model was developed by the Buffalo-Red Watershed District to ensure that the project would not cause an increase in water depths in the upstream legal ditch.

The project entailed excavation of 5.5 kilometers of sinuous, meandering stream channel through native prairie in Atherton State Wildlife Management Area. This was also an area identified by the Minnesota County Biological Survey as a site of biological significance because it contained state-listed species of special concern such as small white lady's-slipper (*Cypripedium candidum*) and northern gentian (*Gentianella amarella*). Conversely, some areas, especially along State Ditch 14, were dominated by invasive reed canary grass (*Phalaris ardundinacea*) and hybrid cattail (*Typha × glauca*). To minimize impacts to native prairie, excavation was done in the winter. Snow and plant litter were removed from the footprint of the new channel to allow the ground to freeze and support equipment. To further lessen the impact, excavators tracked over the channel footprint as they dug the channel and haul roads were minimized. Fill from the excavated channel was used to plug State Ditch 14 and create a series of wetlands. Where the restored channel passed through monocultures of reed canary grass or hybrid cattail, fill was sidecast, spread, and seeded with native grasses (family Poaceae) and wildflowers.

Several challenges were encountered associated with restoring a stream within a legal ditch system. Channelized streams that are part of the legal ditch system are assumed to provide drainage benefits to adjacent farmland by increasing hydraulic conveyance, and physical modifications are generally required to maintain drainage capacity. Traditional design of ditches focused on preventing flooding of agricultural fields by separating connections between the streams and their floodplains by dredging channels deeper or wider. In contrast, stream restoration seeks to reconnect streams to their floodplains. The Buffalo-Red Watershed District played a critical role in working through the legal issues involved with working on a ditch system. Ultimately, the restoration was conducted as a diversion for public benefit, which is a provision within Minnesota drainage laws. The Wilkin County Highway Department, Trout Unlimited, and several landowners also were partners in making the project possible. State Ditch 14 was blocked with a series of plugs that created wetlands within the ditch. County Ditch 40 still remains (Figure 12), but flows were diverted into the restored channel.

Like the reference sites, the restored channel was designed so it would be sinuous, narrow, and deep, with important habitat features (Figure 13). It was designed to be intermediate in dimensions between the upstream (3-meter bankfull width) and downstream (4.3-meter bankfull width) reference channels based on associated drainage area, hydrology, slope, and the landscape. The symmetrical riffle cross sections were 3.7 meters wide and 0.8 meters deep, while the pools were asymmetrical and 1.5 meters deep. Larger, 2.1-meter-deep hammerhead pools, which are formed by meanders having tight radii of curvature, were also incorporated into the design. These naturally formed deep pool features occurred in both reference sites and are key habitats for brook trout (*Salvelinus fontinalis*) and other species. The outside banks of the hammerhead pools were protected with layers of branches and logs covered with fill and native prairie sod mats to provide cover for fish, amphibians, and mammals. Since the stream channel would be constructed within silt, sand, and fine gravel associated with glacial Lake Agassiz, fieldstone riffles were built to provide grade control and habitat until natural coarsening of the substrates and growth of submergent vegetation could stabilize the stream bottom.

The project also reestablished riparian wetlands, including one that was 0.4 square kilometers. These provide floodwater storage and habitat for waterfowl along with water quality benefits. The water quality benefits of

FIGURE 12. County Ditch 40, also known as Lawndale Creek, prior to restoration.

a connected floodplain and wetland were illustrated during a flood event in June 2014. Floodwater flowing out of the stream into the floodplain was turbid (21 NTU), while flow was clear (<5 NTU) as it reentered the stream. Fine sediments could be seen as they collected on the riparian vegetation, and turbidity levels decreased as flows filtered through the floodplain.

The biological response to the restoration has been substantial since the reconnection of the restored channel in August 2011. As of September 2015, thirty species of native fish have been identified in the restored reach of Lawndale Creek. These include several regionally rare headwater species (i.e., northern pearl dace [*Margariscus nachtriebi*], river darter [*Percina shumardi*]) and many common species. Brook trout were marked by fin clipping and stocked in the stream post-restoration by Minnesota Department of Natural Resources staff. While native to Minnesota, brook trout were historically introduced to Lawndale Creek in the 1950s. Several years of major floods prior to the restoration apparently resulted in poor brook trout reproduction, as none were collected in the spring and summer of 2010. Natural reproduction of brook trout (unclipped parr) was not

FIGURE 13. The restored Lawndale Creek following restoration. Restoration actions included re-meandering and narrowing the channel as well as revegetating the banks and floodplain with native prairie vegetation.

documented until 2014. Apparent northern pike (*Esox lucius*) reproduction, indicated by the presence of young of the year northern pike, was also documented in the restored reach and the upstream reference reach, presumably in the connected riparian wetlands where these fish spawn. Thirteen fish species were documented from County Ditch 40 from four surveys conducted between 2004 and 2010. In contrast, twenty-seven fish species were documented from the upper restoration reach and twenty-four fish species were documented from the lower restoration reach in five surveys conducted from 2011 to 2015. Differences in water temperature and geology likely affected the species composition in the two restoration reaches. Summer water temperatures (June 1 to August 31) averaged 15°C in the upper restoration reach and 18.6°C in the lower restoration reach. This may explain the addition of warmwater species such as bluegill (*Lepomis macrochirus*) and rock bass (*Ambloplites rupestris*) in the lower reach. During the same time, little change in the number of species was observed

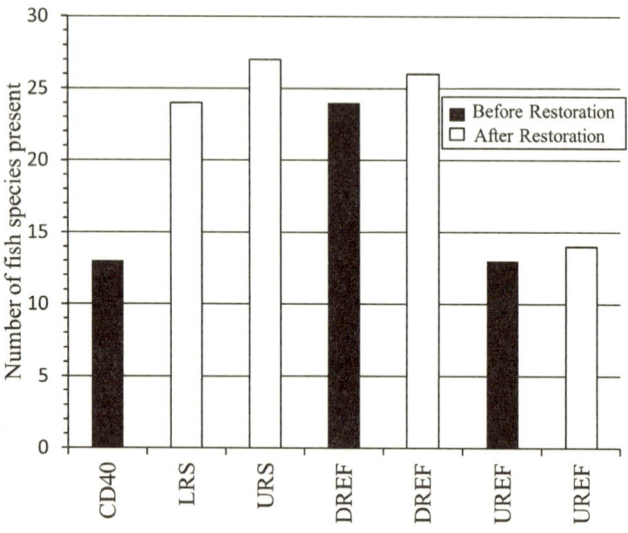

FIGURE 14. Number of species present at Lawndale Creek fish sampling sites before restoration (black bars, 2004–2010) and after restoration (white bars, 2011–2015). Site abbreviations are CD40: County Ditch 40; LRS: lower restoration site; URS: upper restoration site; DREF: downstream reference site; UREF: upstream reference sites. The figure also includes data collected by the Bell Museum in 1975 and the Minnesota Pollution Control Agency in 2009 from the reference reaches.

in the upstream and downstream reference sites, with the upstream site having thirteen species prior to restoration and fourteen post-restoration and the downstream site having twenty-four species prior to restoration and twenty-six species post-restoration (Figure 14).

Other aquatic, semiaquatic, and terrestrial wildlife also benefited from the project. River otter (*Lontra canadensis*), mink (*Neovison vison*), muskrat (*Ondatra zibethicus*), beaver (*Castor canadensis*), and northern water shrew (*Sorex palustris*) were observed in the restored channel along with waterfowl, amphibians, and reptiles. Benthic aquatic macroinvertebrates colonized quickly, but their taxa composition changed with time. Initially, taxa were dominated by chironomids (family Chironomidae), blackfly larvae (family Simuliidae), and wetland species, followed by increasing numbers and diversity of mayflies (order Ephemeroptera), caddisflies (order Trichoptera), and stoneflies (order Plecoptera). Native fatmucket mussels (*Lampsilis siliquoidea*) brought in by migrating host fish began colonizing riffles quickly. Colonization of submergent vegetation, including sedges (family Cyperaceae), bulrush (*Scirpus* spp.), and pondweed (*Potomogeton* spp.), was observed on portions of the streambed as early as July 2011, with more established patches by November 2012. Small white lady's-slippers bloomed in the riparian habitat during the spring after construction in places where excavators and other machinery had driven on the frozen ground.

Remeandering channelized streams in rural areas can restore natural channel geometry, reestablish instream habitat features beneficial to aquatic life, and increase stability and resilience. Many of the most altered stream channels also lie in watersheds altered by cultivation, agricultural drainage, or urbanization. This is true of land use in the Lawndale Creek watershed, which has 73 percent row crop coverage with extensive amounts of newly laid subsurface tile drains. These altered watersheds have increased water yields and accentuated flood events that need to be considered in stream restoration design with broad, connected floodplains that can carry these large events. While restoration of the geomorphology and instream habitat of altered river reaches with natural channel design can improve habitat, stability, and nutrient processes, watershed management practices to improve instream water quality and to reestablish more natural flow regimes will also be needed as part of river restoration in the Midwest.

DAM REMOVAL

As dams across the Midwest age, they are becoming structurally unsound and dam failures are increasingly common (NRCS 2000). Many of the dams' reservoirs have filled with sediment and lost their original functions of water storage and power for mills. There is also an increasing awareness of dam-related deaths due to the hydraulic undertows that occur below many dams. The need to address these issues has led to increasing opportunity for dam removal and river restoration (see Chapter 2).

Dams are among the most important and definitive causes of extinction and extirpation of native freshwater species (Rinne et al. 2005; Aadland

2015) by degrading habitat, blocking spawning migrations, and preventing recolonization following drought, severe winters, water quality–related mortality, and other disturbance events. To address these problems, a growing movement to remove dams for river restoration in the Midwest started in the 1990s, especially in Wisconsin and to some degree in Minnesota and Michigan.

Many dams were built at bedrock rapids and falls capable of providing solid dam foundations and representing a good source of potential power. As a result, many towns that were named for these falls and rapids are now dam sites (e.g., Taylors Falls, Minnesota; Wisconsin Rapids, Wisconsin; Cedar Rapids, Iowa; Grand Rapids, Michigan; Rapids City, Illinois) with the falls and rapids inundated by the dams. Frequently, dams were built as sawmills or gristmills and later for hydroelectric power, but structural deficiencies and sedimentation eventually made them obsolete. These steeper reaches, relatively rare in low-gradient midwestern river systems, were critical spawning habitats for lake sturgeon (*Acipenser fulvescens*), paddlefish (*Polyodon spathula*), walleye (*Sander vitreus*), blue sucker (*Cycleptus elongatus*), and many other native fish that have become rare due to the lack of these habitats.

An assessment of the effects of thirty-two dams on the fish communities of the associated watersheds across Minnesota provides perspective on the magnitude of dam effects on biodiversity (Aadland 2015). All documented, reliable fish collections upstream and downstream of dams were evaluated to determine the fragmentation effects on species richness. This analysis included 150 fish species of which 134 are native to Minnesota. Of the native species found in the associated watershed, an average of 37 percent were absent from collections upstream of the dams, and 41 percent of the species were absent upstream of dams that functioned as complete barriers to fish movement. Pollution-tolerant species including common carp (*Cyprinus carpio*) were least likely to be absent upstream of barriers, while intolerant and imperiled species were most likely to be absent. Habitat differences, stream size, and other factors may contribute to the observed changes in species composition above and below dams. However, many species' ranges were found to end abruptly at dams despite similar habitat upstream. Interestingly, historical records showed that some of the largest fish species, such as lake sturgeon, existed in the upper reaches of some of the smallest watersheds assessed.

Dam removal can result in the dramatic return of native fish to the upstream watershed and also often restores spawning habitat critical to fish populations in downstream rivers and lakes. Extirpations attributable to fragmentation by barriers can be verified when these dams are removed. Removal of ten dams in Minnesota (where adequate post-removal sampling was done) resulted in an average return of 66 percent (23 percent to 89 percent) of the species that had been absent upstream of the dam, including imperiled species such as paddlefish, lake sturgeon, blue sucker, black buffalo (*Ictiobus niger*), and American eel (*Anquilla rostrata*), along with important game species including smallmouth bass (*Micropterus dolomieu*), sauger (*Sander canadensis*), channel catfish (*Ictalurus punctatus*), and flathead catfish (*Pylodictis olivaris*) (Aadland 2015). The return of many fish species often happened shortly after dam removal, but the recovery of rarer fish species is likely to take much longer.

Deep accumulation of fine sediment in reservoirs creates challenges for dam removal (Heinz Center 2002). Sudden release of large volumes of sediment can cause downstream effects to freshwater mussels and in the short term can fill pools and bury coarse substrates. Where sediments are cohesive, head-cutting and incision by the stream can result in an unstable, entrenched channel with a disconnected floodplain until a new floodplain is formed. It is important to consider that these accumulated sediments above the dams would have naturally been transported downstream were it not for interception by the dam and that sediment is an important part of the stream ecosystem. The interception of sediment above dams has also led to the downcutting of streambeds below the dams. Dam removal helps to address the sedimentation and incision problems caused by dams. Releasing some sediment during dam removal may benefit downstream channels. Higher flows were purposely released from Glen Canyon Dam into the Colorado River in 2013 to initiate sediment deposition on sandbars and remediate habitat lost due to nearly five meters of dam-induced bed incision (Grams et al. 2015). Conversely, sudden release of large amounts of accumulated sediment associated with dam removal can overwhelm the downstream channel and cause short-term impacts to aquatic habitat and biota (Aadland 2010). In the next two subsections, we discuss dam removal without active stream channel restoration and dam removal that includes stabilizing sediments by restoring natural channel morphology in the former reservoir.

Dam Removal without Active Restoration and
Management: Vegetation and Sediment Issues

Dam removal presents the opportunity to fully or partly restore many river processes that had been impaired. Channels rapidly downcut and provide passage for fish species almost immediately upon dam removal. Yet geomorphic recovery and other ecological aspects can take longer to recover without active management (Shafroth et al. 2002). Riparian plant communities in particular require years or decades for recovery, particularly to establish mature forest communities. Development of plant communities in drained impoundments after dam removal is influenced by the site conditions, the availability of plant colonizers from surrounding areas, and the competitive influence exerted by the initial plant colonizers over conditions for establishment. If highly aggressive or allelopathic plants colonize the bare mudflats after dam removal, it can delay the establishment of the desired plant community. For these reasons, vegetation is sometimes managed vigorously by planting seed mixes or seedlings to accelerate plant community succession to some endpoint that is more desirable for ecological reasons, human usage of the site, or reduction of sediment loss downstream.

Five dam removal sites in south central Wisconsin that were taken offline between 1960 and 1996 were evaluated in detail in the late 1990s and again by aerial photos in 2015 to examine the development of plant communities in drained millponds over decades following dam removal. Two of the dams were removed from small trout-supporting streams (Mount Vernon and Black Earth Creeks) in the late 1950s through the 1960s. The other three sites, located on the Bark, Maunesha, and Yahara Rivers, were removed for safety and economic reasons in the 1990s. There was widely variable initial colonization of vegetation types and different long-term outcomes that were influenced by the initial conditions after dam removal, sedimentation history, adjacent plant communities, and management during the post-dam removal period (Lenhart 2000).

All five sites had large residual sediment deposits in the drained impoundments, with average depths ranging from 0.3 to 2.3 meters deep and consisting of silty to sandy loam material. The substantial sediment deposits strongly influenced post-dam removal hydrology and plant community development. Hydrologic conditions varied as well, but four of the five sites were highly entrenched, reducing floodplain connectivity and ultimately leading to plant communities more characteristic of uplands. The flood-plains on the drier sites, located along the Yahara and Bark Rivers, Black Earth Creek, and Mount Vernon Creek, were found to be inundated infre-quently, only at flow levels exceeding the six- to thirty-one-year recurrence interval flow. At the Maunesha River site, there was much less sediment accumulation and the post-dam removal floodplain was inundated by the annual or two-year flood.

While overbank flooding strongly influences plant community types, in some situations groundwater inputs may support wetland vegetation. The Bark River site in particular had consistent groundwater discharge inputs to the floodplain (i.e., groundwater upwelling occurring in 80–100 percent of well readings), and it thus supported a higher percentage of wetland plant species than the other sites.

The initial plant colonizers, which were determined to originate from airborne dispersal rather than seed bank, strongly influenced plant commu-nity composition and the establishment of woody plants for years following dam removal (Lenhart 2000). At the sites where dams were removed in the 1990s and surveyed shortly after, rice cutgrass (*Leersia oryzoides*) and reed canary grass were most abundant on the wetter sites, while stinging nettle (*Urtica dioica*) dominated in drier locations and occurred in 80 percent of the sampling plots at the Yahara and Bark River sites. These dominant species inhibited and delayed the establishment of more desirable native plant communities. From 1998 to 1999, woody plant coverage ranged from 0 to 25 percent of the former impoundment area at all five sites (Lenhart 2000). Based on a 2015 aerial photo assessment of the same sites, woody plant coverage had greatly expanded on some sites, with coverage ranging from 15 to 90 percent in riparian areas that were determined to have been forested prior to dam construction. However, only in the Maunesha River site has a woody plant community reestablished itself over the majority of the former millpond area. Here, the plant community still exhibits the shrub swamp stage, with sandbar willow (*Salix interior*) covering much of the area instead of typical mature floodplain forest species.

The sites where dams were removed more than fifty years ago, at Mount Vernon and Black Earth Creeks, were considered successful since they im-proved instream conditions for native brook trout and brown trout (*Salmo trutta*). These projects did not restore native plant communities, as they were subsequently managed as a hayfield (Mount Vernon Creek) and a city park (Black Earth Creek). More recently, a remeandering project was completed on a portion of Black Earth Creek with a small native prairie installed alongside it. In contrast, the three sites removed in the 1990s, on the Bark, Maunesha, and Yahara Rivers, had a mixture of management successes and problems. They were successful in that they reestablished the free flow of water and sediment while improving connectivity for aquatic life. However, native forest communities had still not been reestablished as of 2015 in areas where forest was the dominant plant community type. This is perhaps not unexpected since it is well known that riparian forest ecosystems take many decades to be restored to a composition and structure resembling a mature forest when starting from scratch (Sauer 1998). In contrast, streams with riparian habitats dominated by herbaceous prairie plants can recover more quickly, in a matter of a few years in terms of plant coverage and biomass rather than decades.

From a water quality standpoint, erosion of the steep banks that formed as the river cut through deep reservoir sediments has been a management issue. The Bark, Maunesha, and Yahara River sites, examined shortly after dam removal, had substantial riverbed and bank erosion that occurred as the channel downcut through former sediments, leaving steep banks up to three or four meters high. In order to circumvent many of these sediment and vegetation management issues, some dam removal sites have been more actively managed. For example, the Appleton, Minnesota, case study discussed in the next subsection demonstrates this approach.

Many lessons were learned from early dam removal projects in the 1990s. Today, sediment management and natural channel design techniques have been used in many drained reservoirs to accelerate the recovery process following dam removal (Aadland 2010). By developing a design that uses the accumulated sediment left behind in the reservoir as the new floodplain elevation, engineers and scientists who designed these dam removal projects have excavated meandering channels that are both stable and provide diverse habitat. Since accumulated reservoir sediments are often much finer and less dense than natural streambed materials, they are also very prone to channel incision (Doyle et al. 2005; Aadland 2010) and require specific consideration in design.

Seeding of the floodplain can accelerate native plant community establishment and reduce sediment movement. This is the approach often used today in dam removal projects throughout the Midwest and the United States more widely (Chenoweth et al. 2011). However, the plant communities that will prevail over time in these environments are likely to be composed of species that prefer dry conditions due to the higher elevation of the floodplain following post-dam removal channel incision. Consequently, the use of reference floodplain ecosystems as a guide for dam removal sites may be inappropriate, as dam removal sites with substantial sediment deposits are likely to support fewer hydrophytic plant species than plant communities found on floodplains uninfluenced by dams.

Dam Removal with Active Restoration of the Stream Channel

The Appleton Milldam, located in Appleton, Minnesota, was built in 1872 and was originally 4.9 meters high with flashboards. Over the years, it filled with fine sediments up to 4.6 meters deep. The reservoir was shallow and dominated by common carp and black bullheads (*Amerius melas*). Seventeen native fish species and four native freshwater mussel species found in the Pomme de Terre watershed were absent in the watershed upstream of the dam. Lake sturgeon and several other native fish were known to have existed upstream of the dam based on early records or archaeological digs of campsites (Mulholland et al. 2011). Walleye were captured below the dam but were not found in the reservoir upstream of the dam.

The dam became structurally unsound, and in 1994 the city of Appleton passed a resolution to examine all options to remove or improve the dam. The subsequent study was supported by a $50,000 award from the Minnesota legislature to the city. On January 9, 1997, after considering costs of rebuilding the dam, historic values of the reservoir that had been largely lost by sedimentation, and environmental issues, the city decided to remove the dam rather than rebuild it. In April 1997, the dam failed during record flooding and partially drained the reservoir through a breach under the dam crest. The dam was partially removed to further draw down the reservoir in July 1998 during low summer flows. By this time, floodplain tree species such as green ash (*Fraxinus pennsylvanica*), silver maple (*Acer saccharinum*), and other riparian vegetation had already begun to colonize the sediments exposed by the dam failure. The dam was entirely removed in March 1999 and replaced by 0.6-meter-high rock arch rapids to provide grade control and stabilize reservoir sediments. Two fieldstone riffles were subsequently constructed at inflection points upstream of the rapids in January 2000 to

FIGURE 15. The Appleton dam site during removal in 1998 (A) and after removal, restoration, and revegetation in 2012 (B).

provide additional grade control, habitat, and assure floodplain connectivity (Figures 15 and 16).

Channel design was based on geomorphic surveys of several natural reference reaches upstream of the reservoir and an additional survey at the USGS gauge downstream of the dam. The reference channels were generally associated with a healthy riparian zone and were chosen based on stability, instream habitat diversity, and the presence of a connected floodplain. The geomorphic surveys were conducted during a bankfull discharge event. While this was difficult due to associated depths and velocities, it enabled empirical determination of bankfull discharge. The downstream gauge allowed cross-referencing bankfull discharge, as well as determining the recurrence interval of the bankfull event. Meander pattern geometry was determined from aerial photography.

The channel design needed to incorporate reference channel geometry that adjusted for differences in slope and hydrology and fit within the former boundaries of the upstream reservoir. More important, the channel design needed to be acceptable to adjacent landowners and the community. This was complicated by landownership and legal abstracts that used the edge of the water as a boundary. After several iterations, a key landowner donated land to the city, which enabled the current channel pattern. The project was designed as two reaches with a steeper downstream reach and a lower-slope upstream reach to match the grade of the accumulated reservoir sediments. The upstream channel riffle dimensions were 20 meters wide and 1.25 meters deep with 2:1 side slopes, while pools were 2.16 meters deep with 4:1 slopes on the point bar and 2:1 on the outside bank on a 0.03 percent slope. The downstream channel had similar depths and side slopes but was 17.4 meters wide on a 0.08 percent slope. Outside banks were protected with tree root wads driven into the banks, erosion-control fabric, and willow (*Salix* spp.) staking. Eight arch-shaped fieldstone riffles were constructed for grade control and habitat, and placed at inflection points of the meanders. These arch-shaped fieldstone riffles have been applied at numerous restoration projects in Minnesota and create convergent flow conditions that reduce hydraulic stress on the banks while maintaining deep pools and riffle habitat for spawning fish and benthic aquatic macroinvertebrates.

Excavation and restoration of the river channel was conducted in January and February 2001 during low, stable flows. Unfortunately, spring brought a near-record snowmelt flood that inundated the entire floodplain for sev-

FIGURE 16. Appleton reservoir as it appeared in 1997 (A), after removal of the dam in 2000 (B), and after river restoration and twelve years of tree growth in 2013 (C).

eral months. Since none of the native seeds or willow stakes had time to sprout prior to the flood, the site was vulnerable and the fill used to plug the straight pre-project channel eroded. However, the restored channel experienced little damage, and repairs were made quickly. Riparian trees grew quickly in the following years and have established a floodplain forest.

Since removal of the dam, nine fish species that were absent above the dam have returned to the upstream river reach that stretches for seventy-two kilometers until the next dam at Morris, Minnesota. These include channel catfish, white bass (*Morone chrysops*), freshwater drum (*Aplodinotus grunniens*), emerald shiner (*Notropis atherinoides*), carmine shiner (*Notropis percobromus*), quillback (*Carpoides cyprinus*), silver redhorse (*Moxostoma anisurum*), greater redhorse (*Moxostoma valenciennesi*), and banded darter (*Etheostoma zonale*). Three of the four freshwater mussels that had been extirpated upstream of the dam have recolonized the river upstream to Morris Dam, likely due to the return of their host species. Freshwater drum are the sole host for deertoe mussels (*Truncilla truncate*), and several sucker species (family Catostomidae), including silver and greater redhorse, serve as hosts for elktoe mussels (*Alasmidonta marginata*), so the return of freshwater drum and three sucker species likely explains the return of these two mussel species. Pocketbook mussels (*Lampsillis cardium*) parasitize walleye and several species of sunfishes (family Centrachidae) that were present upstream of the dam. However, if one of the droughts that occurred when the dam was in place had caused these freshwater mussel species to be extirpated, then there would be no adult freshwater mussels and no mussel reproduction, regardless of the presence of the host fish species. The return of these freshwater mussel species following dam removal verifies that the dam was the cause of their extirpation. The restored river in the former Appleton reservoir is now a popular walleye fishery, a species that was not present in the dam reservoir before the restoration.

Similar approaches to dam removal and river restoration have been used at other sites in Minnesota (Aadland 2010). The observed fish community responses to dam removal have been a significant success story. For eleven dams that had adequate post-removal assessments, an average of 66 percent of the fish species that were absent upstream of the dams returned following removal (Aadland 2015). When the Flandreau Dam on the Cottonwood River in southern Minnesota was removed in 1995, twenty-one of twenty-four species that were absent prior to removal returned. This increase in species richness extended across the 369-square-kilometer watershed and roughly 3,200 kilometers of stream, including tributaries. At a number of dam removal sites, rare species have been among the species that returned. Removal of the Minnesota Falls Dam near Granite Falls in 2013 has already ushered in the return of three threatened fish species (paddlefish, blue sucker, black buffalo), one fish species of special concern (lake sturgeon), sauger, flathead catfish, shovelnose sturgeon (*Scaphirhynchus platorynchus*), American eel, silver lamprey (*Ichthyomyzon unicuspis*), highfin carpsucker (*Carpoides velifer*), mooneye (*Hiodon tergisus*), and gizzard shad (*Dorosoma cepedianum*). The restored rapids, previously inundated by the dam, are rare and critical habitat for the lake and shovelnose sturgeon, paddlefish, blue sucker, sauger, and other species. As a result, dam removal can benefit the entire watershed, both in the upstream and downstream direction. Increases in biodiversity following dam removal have also been documented in other parts of the Midwest, such as Wisconsin (Doyle et al. 2005).

CONCLUSIONS

Streams have been highly altered in the Midwest for drainage, hydropower, and direct channelization, while hydrologic alteration and increased sediment and nutrient loading have caused more subtle but equally damaging impacts. While fish habitat structures have been installed for decades in midwestern streams to improve sport fish conditions, ecological restoration of streams to reestablish fluvial, hydrologic, and ecological processes has only been purposefully done in the past twenty years or so. Increases in biodiversity were documented following the restoration of channelized streams and after dam removal. While responses to the restoration of channelized streams are generally confined to the length of the reach restored, dam removal has a watershed-scale effect on biodiversity. Furthermore, restoration of a stream reach is unlikely to have a significant biodiversity response if a downstream barrier prevents immigration. In the western Minnesota case examples (Lawndale Creek and Appleton Milldam), natural reference channels were used to guide design of the newly constructed stream reaches to maximize geomorphic and ecological functions while expediting channel evolution processes. While many of the benefits of these projects were documented shortly after completion, recovery of freshwater mussel communities, populations of long-lived fish species such as sturgeon, native riparian prairie and forest communities, and complex ecosystem components and their interaction can take decades or longer to recover, as demonstrated in the Wisconsin case studies.

Future water quality management efforts for the purpose of addressing total maximum daily loads (TMDLs) and other water quality issues will require a holistic examination of all components of streams in the Midwest. In most midwestern streams, changes in watershed management are needed to restore a more natural flow regime (Poff et al. 2010) and subsequently address changes to sediment, nitrogen, and phosphorus loads (see Chapter 9), especially since many streams in the northern and western parts of the Midwest have experienced large stream flow increases in recent decades.

Changing climate and land use will both continue to make ecological restoration efforts in streams more challenging. Climate change has led to increased and more intense rainfall and streamflow in the Midwest (Lenhart et al. 2011; see Chapter 7). Consequently, the need to restore a hydrologic regime more favorable to native aquatic life is essential for ecological restoration. Land use continues to evolve, with growing urban areas and shifting agricultural practices. In this evolving scenario, the strategies and types of stream restoration projects are likely to evolve as well. Streams with connected floodplains will be more resilient to the extreme hydrologic events associated with climate and land-use change. As in past climate shifts, the suite of species that make up stream communities is likely to change. The resilience of aquatic communities will be dependent on free-flowing streams with unimpeded migration routes that enable the biota to survive extreme hydrologic events and a warming climate. The long-term health of stream ecosystems will depend on restoring this resilience and connectivity.

Like human health, where preventive measures can eliminate the need for corrective surgery, the need for stream restoration can be eliminated by avoiding damages. Some of the conservation lessons learned in the 1930s about riparian buffers, grassed waterways, contour plowing, and wetland preservation have been forgotten with an increased drive toward maximizing production at the expense of sustainability and freshwater ecosystems. A renaissance of conservation practice implementation is needed. Minnesotans voted in 2008 to establish funds through the Clean Water, Land and Legacy Amendment to support increased use of conservation and restoration projects.

Dam construction and poorly designed road crossings, channelization, hard armoring, and other traditional management practices continue to cause substantial and costly damages to water quality, infrastructure, habitat, fisheries, and river ecosystems. Stream restoration has been shown to be a means of repairing these damages and, if done properly, is reliant on natural process and self-sustaining. Ultimately, the health of stream ecosystems will depend on avoiding and minimizing future damages through holistic land and water stewardship.

CHAPTER SIX

RESTORATION OF URBAN ECOSYSTEMS

JEN LYNDALL, JOE DIMISA,
AND CONSTANCE HAUSMAN

INTRODUCTION

Urban environments are human-dominated ecosystems with historical legacies that present unique opportunities and challenges for restoration efforts. The vast majority of urban areas have experienced some level of physical, chemical, or biological degradation due to land-use changes over time. Development and habitat fragmentation have reduced the ecological value of each habitat parcel because they no longer function as a larger, connected ecosystem (Laurance and Yensen 1991). Futhermore, many natural features have been completely eliminated during the process of urbanization. For example, in large cities many small streams have been moved underground into pipes and are now a part of a stormwater drainage network. The challenge for restoration practitioners is to understand which environmental stressors are compromising the function of urban ecosystems. Initial ecosystem evaluations could be based on restoration thresholds related to biotic or abiotic interactions (Whisenant 1999) or landscape scales related to biotic connectivity or changes to physical processes that compromise habitat function (Hobbs and Harris 2001). Localized, targeted management projects (e.g., invasive species removal) may only treat a symptom of the functional change to the ecosystem. Urban environments experience regular disturbances (e.g., development, heat island effects, excessive human use), and as a result restoration efforts need to plan for long-term resiliency (SER 2004). It is especially important to avoid assuming that restoration projects have an attainable historical reference condition and that a discrete effort is sufficient for success (Pickett and Parker 1994). Design and implementation of restoration projects must balance the limitations of urban environments (e.g., land-use constraints, developmental pressure, legacy contamination, invasive species, developed floodplains, high cost of real estate) with the decree that we must start small and obtain public support to create momentum for larger efforts capable of providing greater ecological benefits.

These urban land-use limitations may be further complicated by human dynamics. People are highly integrated into the urban environment and play significant roles in restoration activities as property owners, community members, vested stakeholders, volunteers, advocates, and opponents. As such, the public's level of understanding, acceptance, and assistance can significantly contribute to the success of urban restoration projects (Hobbs 2007). At the same time, the loss of population in formerly densely populated urban cores of midwestern cities has created new needs and opportunities to restore native vegetation and ecological functions.

In this chapter, our objectives are to demonstrate the importance of integrating humans as part of the ecological restoration process in urban ecosystems and to highlight two Midwest case studies that illustrate urban restoration of river, wetland, riparian, and upland ecosystems.

MIDWEST CITIES

Urban areas in the Midwest present excellent opportunities and unique challenges for restoration. The region has great contrasts in land uses, from vast landscapes of agricultural lands to the wilderness of mature forests to large metropolitan cities and suburbs. Indeed, four of the twenty most populated urban areas in the United States are in the Midwest and include (1) Chicago, Illinois (ranked third, with nearly ten million citizens); (2) Detroit, Michigan (ranked fourteenth, with 4.3 million citizens); (3) Minneapolis–Saint Paul, Minnesota (ranked sixteenth, with 3.5 million citizens); and (4) Saint Louis, Missouri (ranked twentieth, with 2.8 million citizens) (U.S. Census Bureau 2016; Figure 17). The Midwest is also well represented in the second tier of most populated cities in the United States (in the twen-

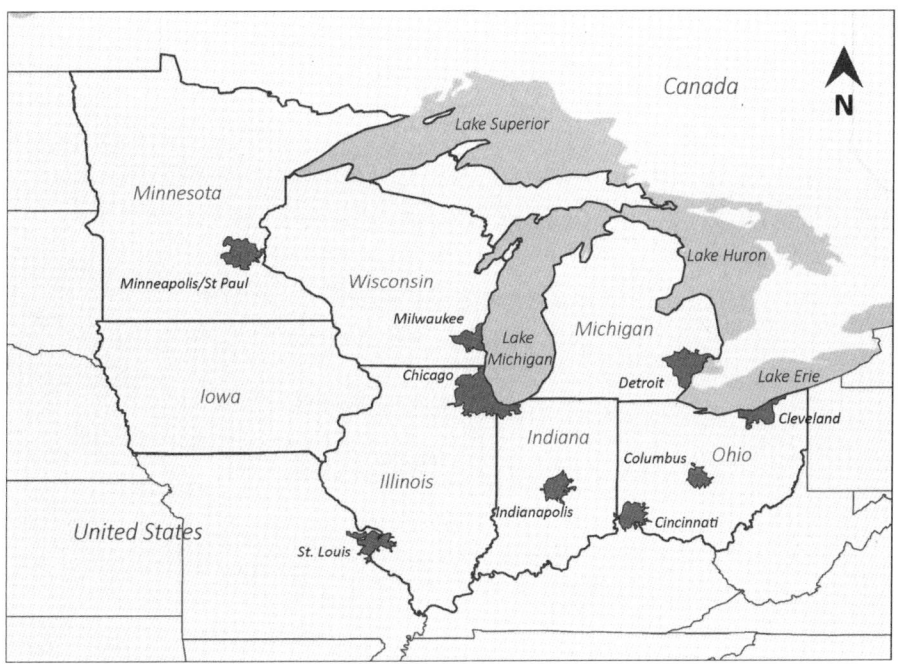

FIGURE 17. Map of the midwestern United States showing locations of large urban areas within the region.

tieth to forty-fifth most populated cities), which includes Cincinnati, Ohio (ranked twenty-eighth), Cleveland, Ohio (ranked thirty-first), Columbus, Ohio (ranked thirty-second), Indianapolis, Indiana (ranked thirty-fourth), and Milwaukee, Wisconsin (ranked thirty-ninth) (U.S. Census Bureau 2016; Figure 17). These five cities are all populated with 1.5 to nearly 2.2 million citizens each (U.S. Census Bureau 2016). Additionally, four of these large midwestern cities (Chicago, Detroit, Cleveland, and Milwaukee) are connected to the Great Lakes, an important regional resource that provides beneficial economic, social, recreational, health, and ecological services. Containing one-fifth of the world's freshwater supply, the Great Lakes provide the region with freshwater-related ecosystem services consisting of drinking water, sport and recreational fisheries, and commercial shipping, which are critical to the vitality of many midwestern states.

Many cities in the Midwest were sited for commerce and established along streams, rivers, wetlands, and the Great Lakes because much of the commerce at their time of establishment depended on water (Ashworth 1986). Ideal sites were those with a good harbor located at established Native American commercial crossroads (Ashworth 1986). These cities, originally linked by water routes and Native American traces, would later develop into river and lake ports, headwater mill towns, and shipping and railroad commerce centers linked by an interconnected system of roads, railroads, canals, and shipping routes. These cities' rapid population growth led to widespread and near-complete removal of existing terrestrial and aquatic ecosystems to facilitate economic development and land settlement.

Midwestern cities started growing following the various treaties with Native Americans, beginning with the Treaty of Greenville, signed in Ohio in 1795. Cities such as Chicago took off as ideal locations for connecting shipping through the Great Lakes with canals and railroads to support settlement of the American West (Cronon 2009). Detroit began as an early French military outpost, grew to be a shipping port, and later became a hub for industry (including the nation's manufacturing center for automobiles). Milwaukee began as a fur-trading post where the Milwaukee River enters Lake Michigan, later becoming an important port city, hub for brewing, and manufacturer of iron and steel. Cleveland originated as Western Reserve land for Connecticut citizens who had their homes destroyed by British troops in the Revolutionary War and as a port town to trade farm goods from inland Ohio with eastern states. Cleveland later became an important industrial city for the production of steel and refining of oil.

The oil slick that caught fire on the Cuyahoga River on June 22, 1969, was a key moment in midwestern environmental history and is widely regarded as the defining start of the modern American environmental movement (Stradling and Stradling 2008). The fire caused $100,000 of damage to two railroad bridges near the Republic Steel mill, but that was only a small part of the fire's story (Stradling and Stradling 2008). The river had caught fire on multiple occasions prior to 1969, and pollution to the river was actually receding in the 1960s. The 1969 fire caught national attention when *Time* magazine published an article about the fire and placed a picture of the burning Cuyahoga River on the magazine's cover. Even more ironically, the cover picture of billowing smoke and flames engulfing a boat was actually from a previous fire in 1952—no pictures of the 1969 fire are known to exist. Additionally, Cleveland mayor Carl Stokes used the event to publicize the problems of pollution on a national level. He and his brother, U.S. representative Louis Stokes, testified before Congress on the need for a greater level of federal regulation. Their testimony and advocacy supported the passage of the federal Clean Water Act in 1972 (Stradling and Stradling 2008).

Since that time, several federal regulations including the Clean Water Act of 1972; Comprehensive Environmental Response, Compensation, and Liability Act of 1980; Resource Conservation and Recovery Act of 1976; and state regulations have led to a regulatory atmosphere that requires remediation of contaminated sites and restoration to compensate for the losses of natural resource services due to contamination.

The International Joint Commission was created in 1909 under the Boundary Waters Treaty to serve as a binational organization to facilitate coordination between the United States and Canada on issues involving the Great Lakes. Under the 2012 Great Lakes Water Quality Agreement, the International Joint Commission designated forty-three Areas of Concern based on significant impairment of beneficial uses that have occurred as a result of human activities at the local level. Each designated Area of Concern was evaluated for fourteen possible beneficial use impairments to assist with focused remediation that would allow the Area of Concern to be removed from the International Joint Commission list (i.e., delisted). The Great Lakes have also been the focus of two additional funding sources that were intended to remediate and restore Areas of Concern. The Great Lakes Legacy Act of 2002 (reauthorized in 2008) provided funding to facilitate the remediation and restoration of Areas of Concern within the Great Lakes. The Great Lakes Restoration Initiative was enacted in 2010 to facilitate restoration and protection of the habitats in the Great Lakes' watersheds, with a focus on remediation, invasive species control, nutrient runoff control, and habitat restoration. These regulations have contributed to the remediation of environmental contamination within the Great Lakes, laying the foundation for subsequent restoration of impaired ecosystems.

THE CHALLENGES

To accommodate midwestern cities' economic and population growth, changes to the natural landscape were inevitable. Urban-related landscape alterations generally fall under one of three interconnected categories: (1) physical, (2) biological, or (3) chemical. Physical modifications consist of filling or draining wetlands, channelization or enclosure of streams within pipe systems, dam installation and levee construction, hardening of banks and shorelines (e.g., sheet piling, armor stone), habitat fragmentation, increase in impervious surfaces adjacent to streams, and modified hydrology, including increased runoff peaks and flashier hydrology (i.e., increased variability in discharge in response to storm events) that contribute to water quality impairments. Biological modifications may result from invasive species, exotic diseases, shifts in community structure consisting predominately of ruderal or tolerant species, or genetic modifications that result from habitat fragmentation and inbreeding. Chemical modifications include point source discharges, National Pollutant Discharge Elimination System permitted releases, combined sewer overflows, and sanitary sewer overflows. Non-point source pollution from agriculture, golf courses, paved roadways, and parking lots also contributes to chemical modifications in urban areas through the discharge of runoff containing fertilizers, pesticides, road salt, and other contaminants.

URBAN RESTORATION OPPORTUNITIES

Opportunities for restoration in urban settings exist despite numerous challenges. There is typically a high degree of institutional support in metropolitan areas, from organizations such as universities, nonprofit organizations, and conservation-based agencies. Additionally, there is a greater abundance of historical data (e.g., land surveys, hydrological surveys, environmental and biological monitoring) in urban settings due to greater interest in evaluating human health and ecological conditions, which often leads to more frequent assessment.

As previously mentioned, urban areas across the Midwest were often developed in close proximity to major rivers and lakes. The economic value of the land adjacent to these waterways is influenced by direct commerce opportunities and by the recreational use and intrinsic value of the water-body. For these reasons, restoration of river and riparian resources are often considered to be a high priority. For example, the Lake Erie restoration in Toledo and the Mississippi River waterfront in Minneapolis are high priorities for economic and recreational reasons.

Today the Midwest Rust Belt offers endless opportunities for restoring degraded or depopulated urban environments. In these cases, restoration in urban areas may simply involve trying to reestablish a native plant community on a site lacking vegetation, or it might attempt to restore ecosystem function for utilitarian purposes rather than focusing on restoration of historic conditions. In other cases (e.g., sites containing forest remnants), it may be possible to restore plant communities to something resembling a historic condition or, at a minimum, a community exhibiting greater ecological integrity than its degraded state.

Generally, urban restoration has a much stronger human component than agricultural or large-scale natural area restoration projects due to the proximity of large populations living near project areas. There are different types of human investment and levels of public involvement depending on the restoration project. Many restoration projects are supported through state or federal funding sources or are part of a regulatory action and have mandated public involvement requirements. For example, these types of projects are often required to initiate a formal public review period in the design phase to identify and address public concerns early on in the process and adjust the design accordingly. Public involvement may also be informal, such as when casual or recreational visitors and users of newly restored sites serve as additional eyes on the ground and provide feedback about project development. Additionally, many urban restoration projects can be enhanced with public outreach components including educational signs, kiosks, and wildlife viewing opportunities. These public outreach components can increase understanding and goodwill by providing informative context about the changes to a project location while limiting recreational access or use in more sensitive ecological areas.

Arguably the most valuable role for people in restoration comes from their engagement in the restoration process through volunteer activities (Miles et al. 1998). Incorporating volunteers to implement ecological restoration projects is popular, particularly in the urban Midwest. In Illinois, almost 161 square kilometers of rare prairie, oak savanna, wetlands, and woodland ecosystems in urban and suburban communities are monitored and managed by volunteers (Miles et al. 1998). Volunteer groups can range from ad hoc single-project groups to formal volunteer groups. Many volunteers are associated with park districts, local watershed associations, and grassroots organizations. Other groups include civic associations and corporations that dedicate themselves to accomplish conservation projects in their community. Boy Scout and Girl Scout troops conduct numerous conservation and restoration projects as well. The most successful volunteer projects are well designed and planned accordingly, with the skills of the volunteers identified a priori so they can just show up and work. Volunteers assisting with urban restoration projects in the Midwest have contributed to hand cutting and removal of invasive plants, planting trees, trail maintenance or building, and pedestrian bridge construction (Miles et al. 1998).

URBAN RESTORATION BY HABITAT TYPES

In the Midwest, past and current restoration efforts encompass all types of aquatic and terrestrial ecosystems (see Introduction), but notably, many early restoration projects in the region originated in urban areas (see Chapters 1 and 3; Gobster 2010). Although many restoration practices and methods used in nonurban environments are also used in urban environments, the population density and high degree of land-use modifications in urban environments limit the types of restoration practices and methods that may be successful in a particular ecosystem type in these areas. General urban restoration approaches for stream, wetland, and upland ecosystems are discussed in the following subsections.

Stream Restoration

Many urban rivers and streams are degraded and exhibit the "urban stream syndrome," which is characterized by modified channel morphology, increased contamination levels, and flashier hydrography compared to pristine streams (Walsh et al. 2005). These impairments need to be removed or addressed prior to or in conjunction with the restoration of the waterway. In larger urban rivers where significant channelization and sheet piling are present, restoration may include the placement of new artificial structures such as river walls or the modification of existing structures. River walls constructed with materials that increase substrate complexity (i.e., boulders or bricks) support greater plant and aquatic macroinvertebrate diversity than river walls constructed with structurally simpler materials (Francis and Hoggart 2008).

Stream restoration may also include hydrologic modifications such as remeandering historically channelized sections, reconnecting the stream with its floodplain (e.g., daylighting the stream [physically uncovering an urban stream that was placed in underground pipes] or adjusting topography), and other hydrologic adjustments to address flow and/or water quality (e.g., weirs, vanes, channel characteristics). Hydrologic restoration is difficult in urban watersheds due to the high degree of historical landscape alteration and high percentage of paved surfaces. As a result, much work related to urban stream restoration involves simply diverting runoff from paved surfaces into grass swales, rain gardens, infiltration basins, or other permeable surfaces to reduce high runoff peaks and protect downstream water quality. One such example is the restoration of Indian Creek in Calumet, Illinois, which was remeandered from a channelized slag-filled ditch to a stream with more defined meanders and instream substrate diversity (Westphal et al. 2010).

Substrate enhancement can be implemented to improve instream biological habitat. Restoration of urban streams in the Midwest typically includes shoreline or bank stabilization, which may include regrading, cribs, geocells, and/or bank armoring. Vegetation may be restored within the streams themselves and the adjacent riparian habitats by removing invasive species and planting native species.

Wetland Restoration

Urban restoration of wetlands is occurring sporadically in the Midwest. The sporadic nature of this effort is primarily due to the lack of available land and high cost of land in urban environments. In rural settings, wetland restoration is easier because land is available and cost for acquisition is lower than in urban areas. In rural situations, subsurface drainage pipes can be disrupted and drainage ditches can be blocked to facilitate conversion of agricultural land to wetlands. In urban environments, large areas of land are already occupied by urban development and often include many small landowners. This situation greatly complicates wetland restoration in urban areas on a widespread scale. Buying out numerous houses to make way for the restoration of wetlands is not practicable in most locales.

However, opportunities do exist for wetland restoration in urban settings. New developments may be required to provide wetland mitigation to offset wetland losses as a part of the development. Additionally, many parks and recreation organizations (e.g., Cleveland Metroparks, Chicago Park District) have undertaken wetland restoration to incorporate natural settings into the parks that they manage. These restored wetlands allow for visitation and nature interpretation, and tend to be less wild than other restored wetlands in rural areas. Often, wetland restoration in urban areas is conducted in conjunction with the restoration of a contaminated brownfield site with legacy contamination that limits potential redevelopment or reuse of the site. Several state and federal programs target the restoration of brownfield sites and offer grants to facilitate their reuse. While multiple owners sometimes are involved, these types of restoration efforts are achieved without disruption to the community. Remediation of these brownfields leads to an overall environmental lift by decontaminating the site and reestablishing natural habitat. Smaller-scale wetlands are also created in urban areas for improvement of water quality. For example, consent decrees by the U.S. Environmental Protection Agency on sewer service entities and voluntary efforts by stormwater utilities interested in providing improved water quality have facilitated wetland restoration in urban environments. Programs promoting rain gardens in residential lots that have small drainage areas can also promote the development of micro-wetlands in these areas. Stormwater management rules that are becoming more stringent in parts of the Midwest are also contributing by requiring detention-basin retrofits. Stormwater management detention basins that were originally designed as dry basins with mowed turf are being retrofitted to include shallow marsh wetland plantings for water quality improvements.

Upland Restoration

Upland areas in urban settings may experience a variety of stressors due to past and current land uses, climate conditions (e.g., heat island effect, soil moisture content), soil nutrient modifications, and impacts due to invasive species (Pickett et al. 2011b). Additionally, upland prairie or forest remnants in urban settings are generally highly fragmented ecosystems isolated from other natural areas. This habitat fragmentation causes the area to be more prone to impairments from heavy foot traffic and increased predation and mortality due to edge effects (e.g., plant mortality due to deer browse pressure, animal mortality due to decreased areas of refuge and increased hunting). Areas with habitat fragmentation also may experience greater impacts caused by invasive species such as common buckthorn (*Rhamnus cathartica*) and Asian bush honeysuckles (*Lonicera* spp.).

Prairie and forest restoration projects in urban settings may require modifications to prepare the soil so that it can support the establishment of native vegetation (Pavao-Zuckerman 2008). Urban development and site construction activities often include scraping and removing the existing topsoil. This topsoil is often sold to recover costs, leaving behind compacted ground that is devoid of organic matter. In some cases, soils may need to be remediated to remove chemical contaminants that are legacies of historical industrial land uses (e.g., brownfields or Superfund sites). Frequently, urban settings are characterized by nutrient-poor soil that limits potential vegetation growth. Restoration projects must address the soil nutrient conditions in order to facilitate vegetation growth (Klaus 2013). Amendments to soil may include fertilizer, compost, biosolids, and amendments to treat pH.

Seed selection is critical to success of prairie and upland restoration projects. Locally collected native seeds are preferred in order to improve the likelihood of success and to preserve local genotypes. If locally collected seeds are not available, then seeds from the same ecoregion would be the next option to ensure that plants are adapted to the region and are genetically appropriate (Plant Conservation Alliance 2015).

Long-term outcomes of urban forest restoration are dependent upon the type of restoration that is implemented. Urban forest restoration projects typically involve planting larger trees to supplement the canopy and understory and to outcompete invasive species. Simmons et al. (2016) found that tree diversity and canopy closure were greater in areas with exotic species removal and native tree planting than in areas with exotic species removal alone. Tree diversity and native tree regeneration were further improved with repeated removal of invasive species (Simmons et al. 2016).

MONITORING AND MAINTENANCE
OF RESTORATION SITES

Ideally, all restoration projects would be monitored to evaluate project success over time and inform the field of ecological restoration. Monitoring may include the comparison of initial conditions at implementation to conditions over time. Examples of monitoring metrics include native vegetation success (e.g., percent cover, species diversity, size, growth), invasive species management (e.g., percent cover of target species), soil stability (e.g., erosion control) and integrity of engineered structures. When monitoring information indicates that the project is not performing as desired, additional management actions can be implemented to address the problem. For example, if monitoring indicates a large percentage of an invasive species within the restored site, then spot herbicide treatment would be implemented to reduce the percentage of invasive species. This use of monitoring to assist with decisions regarding the post-restoration management of restored sites is known as adaptive management, which is an approach that has been successfully used in restoration efforts in the Midwest (see Chapters 3 and 4).

Monitoring and adaptive management are additional challenges in urban settings for many reasons. Often, restoration funding only supports limited monitoring efforts, if it is included at all. This approach may allow short-term restoration success, but it increases the chances of long-term failures as structures deteriorate or invasive species take over the restored area. For example, the project may meet all success criteria in the first year, but in the second year an invasive species encroaches from the unmaintained adjacent property. Without proper monitoring and adaptive management, the restored area may be dominated by invasive species by the third year. These prospects are especially problematic because it often takes up to ten years for native vegetation to establish successfully (Smiley and Rumora 2015), particularly in the harsh soil and environmental conditions often found in urban settings.

Other types of restoration projects have limited capacity for adaptive management because the project decision-makers may not have the authority to modify the original design based on post-restoration conditions. For example, if a certain planted species exhibits a low survival rate at a restoration site, the project manager would likely want to adjust species replanting lists. However, this may not be permissible under certain funding or regulatory mechanisms (e.g., for projects required under regulatory authority), because it is a deviation from the approved project design.

Restoration projects in urban settings are also occasionally tampered with or vandalized by curious or troublesome people. Some common problems are removal of permanent markers, removal or defacement of signs, littering, removal of fencing, damage to engineered structures, and damage to native vegetation. Unfortunately, little can be done to prevent these kinds of problems. Public outreach including educational signage may help by informing visitors about the purpose of the project and expectations for visitors.

Many of the challenges discussed above make urban restoration projects more high-maintenance than projects sited in less disturbed areas. However, two advantages that urban restoration projects have are that volunteers are often more readily available and the ecological benefits are often more pronounced than they are for restoration projects in less disturbed areas. In the next two case studies, we highlight examples of a successful urban river restoration project and a successful urban watershed restoration project conducted within the Rust Belt of northern Ohio.

ASHTABULA RIVER CASE STUDY

Ecosystem function in and adjacent to the Ashtabula River in northeastern Ohio has been limited by land use, development, and legacy contamination. Although much of the upstream portion of the Ashtabula River is located in an agricultural setting, the land adjacent to the downstream portion (3.2 kilometers prior to the confluence with Lake Erie) consists primarily of urban and industrial land uses. Most of the shoreline in this portion of the river has been modified, either through placement of sheet pile or the presence of docks and marinas. Due to decades of manufacturing and shipping activities and contamination from the tributary Fields Brook, the Ashtabula River has been historically contaminated by a number of chemicals, including polychlorinated biphenyls (PCBs), polycyclic aromatic hydrocarbons (PAHs), and other chemicals of concern. Fields Brook was placed on the National Priorities List in 1983 due to chemical contamination from industrial facilities. Fields Brook remediation under the Comprehensive Environmental Response, Compensation, and Liability Act began in 1997 and is still ongoing. However, the associated contamination in the Ashtabula River was addressed separately.

In 1985, the International Joint Commission designated the lower 3.2 kilometers of the Ashtabula River as an Area of Concern due to several beneficial use impairments. Of the fourteen possible beneficial use impairments, six have been documented at the Ashtabula River and consist of (1) restrictions on fish and wildlife consumption; (2) degradation of fish and wildlife populations; (3) fish tumors or other deformities; (4) degradation of benthos; (5) restrictions on dredging activities; and (6) loss of fish and wildlife habitat. In order to improve beneficial uses and delist the river as an Area of Concern, remediation and restoration have been attempted by the collaborative efforts of multiple agencies and private stakeholders.

Source control of ongoing chemical contamination is critical to the success of any river restoration to prevent recontamination of the restored habitat. Therefore, the remediation of legacy sediment contamination was the initial priority. The earlier remediation of contaminated sediment in Fields Brook reduced the likelihood that this tributary was a source of contamination for the Ashtabula River. Contaminated sediment in the Ashtabula River was remediated as part of an approximately $50 million Great Lakes Legacy Act project. The Great Lakes Legacy Act funding is intended to assist with the cleanup of contaminated sediment in the U.S. Areas of Concern and requires a cost-sharing public-private partnership. Partners associated with the Great Lakes Legacy Act project included the Ashtabula City Port Authority, the state of Ohio, U.S. Army Corps of Engineers, U.S. Environmental Protection Agency, and several private companies. The Great Lakes Legacy Act project involved the removal of more than 380,000 cubic meters of contaminated sediment in 2006 and 2007. Sediments were hydraulically dredged, pumped several kilometers through pipelines to a landfill in the Fields Brook area, dewatered, and capped. In 2013, an additional 8,410 cubic meters of sediment were dredged in the Ashtabula North Slip as part of the project to address the last remaining contamination in the Ashtabula River Area of Concern.

Once the contaminated sediment was removed from the river, restoration was ready to begin. Three separate restoration projects were coordinated to address the only remaining natural shoreline within the Area of Concern.

FIGURE 18. Location of restored section of natural shoreline adjacent to the Ashtabula River. (Aerial image obtained from Google Earth.)

These projects were phased to accommodate funding opportunities and construction logistics. The projects were funded under the Great Lakes Legacy Act, Great Lakes Restoration Initiative, and Natural Resource Damages settlement funding.

In May 2010, a restoration project was completed along a portion of the Ashtabula River's natural shoreline (Figure 18). The project was funded through the Great Lakes Legacy Act and provided mitigation credit under the Clean Water Act Section 404 permit for the project. The restoration included the installation of approximately 244 meters of fish habitat shelf and the removal of invasive species and planting native vegetation in adjacent aquatic and upland areas. The project addressed three beneficial use impairments: degradation of fish and wildlife populations, degradation of benthos, and loss of fish and wildlife habitat.

In 2012, the Ohio Environmental Protection Agency completed construction of approximately 427 meters of fish habitat shelves consisting of constructed shallow gravel beds and restoration of adjacent aquatic and upland areas by removing invasive species and planting native vegetation. This project was funded under the Great Lakes Restoration Initiative. The project was slightly upstream of the Great Lakes Legacy Act restoration project that had been completed earlier in the year. Land and access were provided through a settlement agreement with the state of Ohio, and the project was closely coordinated with the completed Great Lakes Legacy Act fish shelf project and the planned Natural Resources Damages restoration project. The Great Lakes Restoration Initiative restoration project addressed the beneficial use impairments of degradation of fish and wildlife populations, degradation of benthos, and loss of fish and wildlife habitat.

The Natural Resource Damage claims were settled in 2012 and included property acquisitions, human use enhancements, and ecological restoration projects. Human use projects included the enhancement of public access, creation of boardwalks and trails with interpretive signs, and the construction of a canoe and kayak launch. Ecological restoration projects were focused on the area in between the Great Lakes Legacy Act and Great Lakes Restoration Initiative projects. The ecological restoration project included removal of invasive species (e.g., *Phragmites* spp., *Ailanthus* spp.), construction of a hydraulic connector and wetlands, revegetation of riparian habitats with native trees and shrubs, and preserving and conserving restoration areas with conservation easements and environmental covenants. Ecological restoration goals were to improve habitat such that the beneficial use impairments may be removed and the Area of Concern can be delisted. Monitoring work is ongoing to evaluate the success of the project and identify if post-restoration management actions are needed.

The coordinated efforts of these three projects were critical to leveraging the benefits of each individual project. Rather than creating isolated ecosystem enhancements, the project coordination magnified the larger-scale enhancement of the area and expedited the delisting of beneficial use impairments. Multiple agencies collected data to support delisting of beneficial use impairments and the eventual delisting of the Ashtabula River Area of Concern. These agencies included Ohio Environmental Protection Agency, Ohio Department of Natural Resources, U.S. Environmental Protection Agency Great Lakes National Program Office, U.S. Environmental Protection Agency Office of Research and Development,

FIGURE 19. Aerial view of the Acacia Reservation and the surrounding commercial and residential developments adjacent to this restoration site.

National Oceanic and Atmospheric Administration, U.S. Fish and Wildlife Service, and the U.S. Geological Survey. As a result of the remediation and restoration actions, three beneficial use impairments (restrictions on fish and wildlife consumption, degradation of fish and wildlife populations, and loss of fish and wildlife habitat) were delisted in 2014. The remaining beneficial use impairments are believed to be related to the legacy sediment contamination that was removed as part of the previous Great Lakes Legacy Act projects conducted on the Ashtabula River. The Ashtabula River stakeholders hope that this Area of Concern may be fully delisted once the ecosystem has had more time to recover.

As a recreational activity, golf and associated country club organizations have been heavily impacted by hard economic times and a shift in preferred pastimes. In 2013, 157 golf courses closed and only 13 new courses opened (National Golf Foundation 2012). This is part of an ongoing trend, with the number of course closures outnumbering those that have opened in each of the last eight years. Acacia Country Club, located in the northeastern Ohio city of Lyndhurst, existed as a private golf course for nearly ninety years but struggled in the recent past and eventually closed in 2012 (Figure 19). This provided a chance to repurpose the site and an opportunity for ecological restoration.

In December 2012, Cleveland Metroparks acquired Acacia Country Club through a generous donation from the Conservation Fund. Deed restrictions in the transfer prevented future golf play and only allowed for passive recreational opportunities. Now known as Acacia Reservation, this 0.63-square-kilometer green space will undergo various restoration initiatives that restore hydrological function and ecological communities to convert the heavily managed golf course into a natural landscape. The property included eighteen holes partitioned by open woodland roughs, a driving range, forested boundaries, three ponds, two partially culverted headwater streams, and the main stem of Euclid Creek. Euclid Creek is a direct tributary to Lake Erie and has a watershed that encompasses 60.3 square kilometers. Acacia Reservation sits at the headwaters of Euclid Creek, and its transfer in ownership and management to Cleveland Metroparks increased the amount of permanently protected green space in the Euclid Creek watershed by 28.5 percent. Approximately 396 meters of Euclid Creek that runs through Acacia Reservation is entrenched with eroding stream banks, resulting in siltation, sedimentation, and aquatic macroinvertebrate and fish habitat loss. Planned restoration goals focus on these issues: the objective is to restore the natural hydrological conditions with associated wetland and riparian habitat and to stabilize the main stem of Euclid Creek.

Successful restoration projects start with behind-the-scenes efforts that require an appreciation of historic land use and knowledge of existing conditions. Acacia Country Club was designed and constructed in 1921 at a time when Cuyahoga County was still largely rural. Post–World War II suburban expansion in greater Cleveland, particularly in the 1970s and 1980s, led to the conversion of most open space to suburban residential neighborhoods with embedded commercial complexes. Today, approximately 93 percent of the Euclid Creek watershed is classified as urban, with 32.6 percent of the basin covered with impervious surface area (Ohio Environmental Protection Agency 2005).

The dominance of urban land use in the Euclid Creek watershed is the biggest factor influencing the hydrology within the watershed. Construction, dams and impoundment, hydro-modification, stream habitat modification, and urban runoff and development are all identified as sources of impairment in the Ohio Environmental Protection Agency Total Maximum Daily Load (TMDL) calculations for the Euclid Creek watershed (Ohio Environmental Protection Agency 2005) and in the *Euclid Creek Watershed Action Plan* (EC WFC 2006). Initial restoration goals were identified to account for these impacts and to incorporate a public park with sustainable management practices. The goals include (1) restore the natural hydrological function; (2) reestablish native forest and wetland communities; (3) develop adaptive management that incorporates scientific research and stewardship; and (4) integrate public use and social reflection to connect people with their environment. In the next five subsections, we describe details related to the different components of this restoration project.

Baseline Assessments

Immediately after the initial purchase of the Acacia Reservation in 2012, there were efforts to promote civic engagement and involvement. Community involvement included holding public meetings, conducting baseline BioBlitz studies (i.e., intensive short surveys where experts and volunteers attempt to identify all species within an area), and establishing academic research partners. With nine suburban communities and 68,000 people living in the Euclid Creek watershed, it was important to host public meetings to inform the community early on about the process of restoration and to manage expectations of its appearance during the process. Two BioBlitz events were held in 2013 to involve the public and to assist with documenting the baseline biodiversity of the park. Twenty-eight volunteers and a dozen regional professionals from Cleveland Metroparks, local universities, and the Cleveland Museum of Natural History completed eight taxonomic surveys of plants, earthworms (suborder Lumbricina), insects, ants (family Formicidae), fish, amphibians, mammals, and birds within the restoration site. Over eighty species of birds were identified during the surveys. Other university partnerships included class field trip projects to document stream structure and condition with the use of the Headwater Habitat Evaluation Index, Headwater Macroinvertebrate Field Evaluation Index, Bank Erosion Hazard Index (BEHI), and independent student research to identify spatial patterns of tree establishment. As part of a senior capstone project, three students from a nearby high school established research plots to measure natural tree regeneration and deer (*Odocoileus virginianus*) browsing pressure. Additional field sampling implemented by natural resource staff at Acacia Reservation included the establishment of forty permanent vegetation research plots that included soil characterization and fixed reference photos; ten automatic water level recorders to monitor groundwater depths; and two automatic flowmeters to track water level, water velocity, and total volume. Soil test borings were also performed as part of an archaeological survey. These combined efforts aided the inventory of baseline conditions for Acacia Reservation and were used during the ecological restoration master planning process that was completed in May 2014 (Biohabitats 2014).

Golf Course Legacy Effects

The construction of the golf course and the intensive management to maintain manicured turf left behind a legacy of soil impacts. Soil profile modifications occurred due to contour shaping and earthmoving during course construction and the addition of topdressing material. Frequent use of machinery to maintain greens and tees compacted the soil. Hydrologic modifications via subsurface structures also represent impacts that occurred due to golf course operations. A network of irrigation lines and subsurface drainage pipes consisting of corrugated plastic or clay tile was present. Residual soil chemistry was impacted by regular applications of herbicides, pesticides, and fungicides. Mercury and lead concentrations are a potential concern for any golf course in operation prior to the 1970s. Proper soil testing may be necessary to determine concentration levels and location prior to major soil disturbance (MDA 2015).

The research conducted during the restoration master planning process and the baseline data collection efforts identified the legacy effects at Acacia.

Because of limitations in earthmoving technology in the 1920s when the golf course was originally constructed, course topography is likely to follow the natural topography of the area. Therefore, the soils are most likely intact; archaeological soil test borers also support this premise. The subsurface drainage pipes consisted mostly of clay tile with some corrugated plastic sections in limited areas that have recently been replaced. Based on basic soil analysis, most of the property consists of partially hydric soils that are phosphorus limited. Testing for heavy metals in soil is expensive, but preliminary results show slightly greater concentrations of heavy metals on the former golf greens than on the fairways. Additional samples are needed to determine if elevated metal concentrations in the former golf greens are consistent and significant. However for now, soil disturbance on golf greens has been avoided.

The condition of mycorrhizal fungi communities may reflect other possible residual soil effects. These soil fungi form mutually beneficial relationships with plant roots, assisting with the acquisition of nutrients such as nitrogen and phosphorous, and protecting the roots from soil pathogens. Over years, fungi build up extensive subsurface connections, and these fungal communities decline in areas treated with herbicide and fungicide. Pathogens and mutualists in the soil may have large effects on establishing trees, but there is little information available about how to restore tree communities successfully in highly modified urban environments where soils have been exposed to long periods of chemical treatments (Maltz and Treseder 2015). Many of northeastern Ohio's native tree species are ectomycorrhizal (e.g., American beech [*Fagus grandiflora*], hickory [*Carya* spp.], oak [*Quercus* spp.]) (Berliner and Torrey 1989; Brundrett 1991) or arbuscular mycorrhizal (i.e., red maple [*Acer rubrum*)]) (Berliner and Torrey 1989), meaning that they depend on fungal mutualists in the soil (Brundrett 2009). In 2014, Cleveland Metroparks, Case Western Reserve University, and Holden Arboretum collaborated to investigate native tree establishment with and without their ectomycorrhizal and arbuscular mycorrhizal fungal counterparts on the former golf course. The results of this research will aid in understanding soil fungal requirements needed during tree establishment and will give insight to some basic knowledge of soil fungal communities in highly disturbed urban environments.

Successional Observations

Summer 2015 marked the third season since the golf course closed. The former fairways were left fallow and provided an opportunity to evaluate the effectiveness of a passive restoration approach of simply allowing succession to occur in the fairways. The cessation of mowing has facilitated the establishment of woody trees. Young oak trees, predominantly confined to the seed rain of their parent tree, germinated and set their first year of growth in 2013. The wind-dispersed maples (*Acer* spp.) from the same year went greater distances and successfully established on some of the fairways. While these trees were a sign of the beginning of reestablishing a forest, the young recruits were subject to browsing by an overabundant urban deer population. In 2014, a study was initiated to evaluate the impacts of deer browsing on tree establishment by putting protective caging around young seedlings. Ten exclosures and ten controls of one square meter each were placed on fairways and roughs where there was ample tree recruitment. All trees within the plot area were counted, measured, and assessed for injury caused by deer browsing. The maples were the hardest hit, and seedlings outside of the fence have not exceeded a height of fifteen centimeters in the last two growing seasons.

Former golf courses in urban environments face an additional pressure that can limit restoration success: the establishment of invasive plant species. Turf grass forms a dense root mat and a thick aboveground thatch that creates competitive barriers for new plant establishment. However, wind-dispersed seed like that of Canada thistle (*Cirsium arvense*) can quite readily establish along the former fairways. Based on observations of other fallow golf courses and discussions with the park managers who inherited them, Canada thistle is one of the earliest colonizers within the former turf grass and can quickly become difficult to control. Knowing this, early detection and targeted chemical control with aminopyralid was used to treat newly established plants to prevent large-scale infestation. However, annual treatments have been necessary to keep Canada thistle from spreading further. Rapid response and isolated treatments minimize non-target effects, and long-term vegetation research plots will be used to monitor plant changes over time.

Restoration Implementation

Cleveland Metroparks is implementing a multiphase implementation process for the restoration of Acacia Reservation. Phase 1 included baseline data collection efforts to guide the prioritization of active restoration initiatives and was completed in 2013 and 2014. Phase 2, completed from fall 2014 to fall 2015, included the conversion of one fairway to a pollinator meadow habitat. Another fairway was reforested with planted trees and shrubs at a rate of 37,000 trees per square kilometer. These newly converted fairways occur along the edge of the park and created a habitat buffer between the boundary of the reservation and adjacent properties. Additional work during this phase, supported by the Conservation Fund, included a combination of breaking or removal of subsurface drainage pipes throughout the property and the establishment of successional wet meadows on about 0.07 square kilometers. Phase 3 began in 2015. The priority is to restore natural hydrological function of Euclid Creek and associated tributaries.

The Ohio Environmental Protection Agency granted Cleveland Metroparks a Water Resource Restoration Sponsor Program grant for program year 2015 for $1.9 million. Additional funding from the U.S. Fish and Wildlife Service ($150,000), Ohio Environmental Protection Agency 319 fund ($200,000), and Ohio Environmental Protection Agency funding through the U.S. Great Lakes Restoration Initiative ($25,000) will further support design and implementation of various restoration projects at Acacia Reservation that will lead to a healthier, more resilient watershed. Phase 3 restoration activities planned include (1) instream habitat restoration and bank stabilization within 396 meters of the main stem of Euclid Creek; (2) restoration of 0.02 square kilometers of adjacent floodplain to increase infiltration capacity and decrease sediment and pollutant loads into Euclid Creek; (3) fill removal followed by daylighting and reestablishment of floodplain connectivity within sixty-one meters of a Euclid Creek tributary; and (4) the restoration of natural hydrology and soil structure through breaking and removal of subsurface drainage pipes to create more than 0.06 square kilometers of wetland swales along existing drainage corridors to increase infiltration capacity and provide additional native habitat in the watershed. Design for these projects commenced early 2016, with implementation occurring in late 2016 through 2017 (Figures 20 and 21).

Future Directions for Acacia Reservation

The headwater streams and wetland depressions historically present had been buried and modified by subsurface drainage pipes installed to manage

FIGURE 20. *(top)* Euclid Creek, a stream located in the Acacia Reservation, prior to restoration activities in fall 2016. The stream was degraded both physically and biologically and suffered from low floodplain connectivity, poor water quality, and lack of native riparian vegetation. FIGURE 21. *(bottom)* Euclid Creek restoration project shortly after construction in April 2017. The project re-created an active floodplain bench to improve nutrient and sediment removal and reestablished native vegetation along the creek.

the surface water on the golf course. Once completed, the Euclid Creek restoration project will reduce non-point source pollution from urban runoff, reduce erosion rates and flooding incidents, and restore hydrology and water quality protection of the main stem and its headwaters. In its entirety, the restoration of Acacia Reservation is expected to improve stream habitat and water quality on site and will be one of the largest single urban watershed restoration projects undertaken in Cuyahoga County, Ohio, to date. Fairways, tees, greens, and bunkers are being replaced with floodplains, wetlands, meadows, and forest habitats that will ensure the long-term resiliency of this urban green space.

Cleveland Metroparks values the cost-effective ecological restoration and transition of the property to a more biologically diverse and environmentally sustainable landscape. This transition will mature over time and will be studied by a collaboration of institutions assisted by park patrons. Integrating the surrounding community fosters interpretive opportunities for understanding the ecological role of restoration. With each new seedling planted, meadow installed, and stream bank stabilized, park patrons have the historic opportunity to witness the development and growth of restoration in practice. It is intended that the lessons learned and results documented at Acacia Reservation serve as a model for other landowners in urban watershed restoration and golf course renaturalization.

CONCLUSIONS

Urban areas in the Midwest present excellent opportunities and unique challenges for restoration. These areas were originally settled during a period of robust westward expansion in the 1800s, experienced the Industrial Revolution, and later expanded their suburbs during the postwar building boom of the twentieth century. The region currently represents some of the most highly populated areas in the United States. The legacy of population growth left in its wake contaminated land and waters, natural areas degraded by invasive species, filling and draining of wetlands, destruction of native prairies and woodlands, and degraded streams resulting from uncontrolled urban runoff. However, with an optimistic perspective these problems are viewed as opportunities for the urban ecological restorationist.

As evidenced first by the Stokes brothers in Cleveland promoting passage of the Clean Water Act, a great advantage to urban ecological restoration is the opportunity to capitalize on the large numbers of people living, working, and recreating in the vicinity of proposed projects. These people are an excellent source of volunteer assistance, as well as vocal proponents and sources of funding. People who recreate, view, protect, and advocate for urban ecological restoration projects are critical to the success of these restoration efforts.

And yet, many challenges remain. The legacy of urban runoff poses one of the main challenges to urban ecological restoration. Because of this, stream restoration efforts must be supported by watershed management efforts. Wetlands tend to be rare in urban environments, so wetland restoration efforts typically focus on protecting and conserving these habitats, limiting invasive species, and restoring hydrologic conditions. Upland restoration can greatly improve biodiversity in urban settings by focusing on native vegetation and invasive species control in prairies and forests, as well as prescribed fires in prairie and forest habitats. Monitoring and maintenance are critical in urban restoration projects because of adjacent urban influences such as invasive species, pollution runoff, and stormwater inputs. Also, urban restoration projects need to account for human usage, maintenance, and vandalism to ensure success. As highlighted by the case studies, most urban ecological restoration projects do not consist of just one ecosystem type; they often involve a comprehensive effort to restore multiple ecosystem types. Online resources are available to those interested in urban restoration who anticipate needing assistance and information on restoration topics that lie outside their areas of expertise (Table 2).

Urban ecological restoration in the Midwest can be expensive, complicated, and require a great level of stakeholder involvement. However, reversing environmental degradation that can be decades old can be very rewarding. Furthermore, the potential to reestablish a natural environment for a large number of people in urban environments, including some who have never before had access to nature, is an excellent and exciting benefit of restoring urban ecosystems.

TABLE 2. Websites of selected organizations involved with restoration of urban ecosystems in the Midwest and websites with useful tools that can assist with urban restoration efforts.

Organizations	
Chicago Wilderness	http://www.chicagowilderness.org
Global Restoration Network	http://www.globalrestorationnetwork.org
Great Lakes Phragmites Collaborative	http://greatlakesphragmites.net
Great River Greening	http://www.greatrivergreening.org
Lake Erie Allegheny Partnership for Biodiversity	http://www.leapbio.org
Midwest Invasive Plant Network	http://www.mipn.org
Society for Ecological Restoration Midwest–Great Lakes Chapter	http://chapter.ser.org/midwestgreatlakes
The Greening of Detroit	http://www.greeningofdetroit.com
The Stewardship Network	https://www.stewardshipnetwork.org
Wildlife Habitat Council	http://www.wildlifehc.org
Useful Tools	
i-Tree software suite	http://www.itreetools.org
United States Environmental Protection Agency National Stormwater Calculator	http://www.epa.gov/water-research/national-stormwater-calculator
Urban Tree Canopy Assessment	http://www.nrs.fs.fed.us/urban/utc

PART

THREE

The Future of

Ecological Restoration

in the Midwest

CHAPTER SEVEN

THE ROLE OF RESTORATION
IN A CHANGING WORLD

Increasing Ecosystem Resilience and Response

in the Face of Climate Change

JOHN SHUEY AND HUA CHEN

Conceptually, the practice of ecological restoration is simple. Ecological restoration is an intentional activity that initiates or accelerates the recovery of a degraded ecosystem with respect to its health, integrity, and sustainability (SER 2004). Successful restorations define a priori the factors that negatively impact ecosystem health, integrity, and sustainability and develop and implement strategies that alleviate those stressors. Successful restoration programs also develop measurable restoration outcomes that assess status change in those stressors such that restoration strategies can be adjusted to reflect ecological response. By defining ecologically meaningful strategies, outcomes, and measures, successful programs ensure that they understand and can commit the resources required for success. The goal of ecological restoration is to reestablish ecological processes and communities that self-organize into functional, resilient ecosystems that adapt to changing conditions (Milly et al. 2008; Alexander et al. 2011). This goal requires recognizing how ecological processes may differ under future climatic regimes and designing restorations that have enhanced resilience under future conditions.

Recently, Staudinger et al. (2015) developed a detailed synthesis of climate change impacts to the Midwest that is concisely summarized in the following sentences. The Midwest is currently experiencing altered climate, the result of human-induced global climatic warming (Figure 22). Warm-

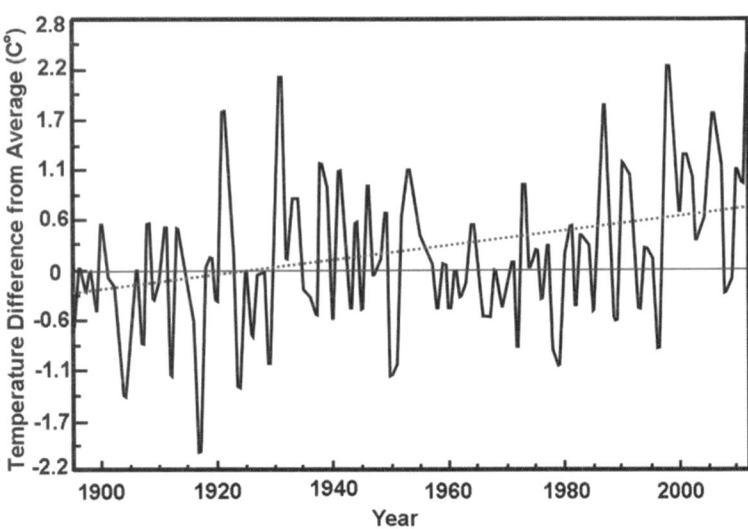

FIGURE 22. Annual average temperatures across the Midwest are increasing. The trend line (dashed line), calculated over the period 1895 to 2012, is equal to an increase of 1.5°F. (Adapted from Kunkel et al. 2013.)

ing is occurring in all seasons, particularly during the winter and at higher latitudes. Winters are getting wetter, with snow shifting to rain, resulting in less snowpack in all areas except downwind of the Great Lakes, where warming enhances lake-effect precipitation. Seasonal minimum winter temperatures are increasing. Earlier thawing of lakes prolongs the period of warming in lakes, promoting more algal blooms. In summer, rainfall events are becoming more intense and occurring less often, which results in minimal net change in annual precipitation amounts (Figure 23). These changes are expected subsequently to influence lake levels, hydrological flows, storm frequency, distributional shifts in vegetation, and ultimately, ecosystem structure and function.

Climate change will have dramatic impacts across global biomes, impacting ecosystem services and human well-being (Mooney et al. 2009; Stein et al. 2013; Pryor et al. 2014). Increasingly, ecological restoration is viewed as a global strategy to partially offset climate change by enhancing carbon sequestration and practices that slow the decomposition rate of organic matter and increase global resilience. There are broad calls for creating policies that encourage such actions worldwide at spatial scales that impact global carbon budgets or regional landscapes (Chazdon 2008; Lal et al. 2007; Trabucco et al. 2008; Zomer et al. 2008). While these global calls for action may lead to increases in restoration efforts, they offer little insight into site-based adaptive criteria that could be incorporated into actual restoration designs.

A seasoned restoration ecologist or conservation biologist can envision how these broad climatic changes will potentially impact ecosystems at the site and regional level. While local or site-based restoration can only incrementally mitigate the underlying drivers and regional impacts of climate change, a well-conceived restoration project can directly address local climate-induced ecological stressors. Those designing restoration projects simply need to connect the dots between current conditions, desired outcomes, and the future predicted climatic conditions to design and implement appropriate restorations. Toward that end, natural resource managers have developed conceptual frameworks for planning and implementing adaptation strategies (Heller and Zavaleta 2008; Moser and Ekstrom 2010). These conceptual frameworks fall into four general approaches applicable to ecological restoration (Millar et al. 2007) that are summarized below.

Increase resistance to change – Manage existing ecosystems and resources so that they are better able to resist the influence of climate change or forestall undesired effects of change for as long as possible. This is essentially an attempt to defy predictable impacts from future climatic regimes, but it may be a defensible approach for certain high-value ecosystems. For example, maintaining critical habitat for endangered species may be a viable strategy to buy time until more adaptive strategies are implemented. Likewise, high-value ecosystems such as production forests or coldwater streams may justify the resources required for resistance. Resisting climatic and other environmental changes typically requires intensive interventions that increase over time. These strategies have a high likelihood of failure over time as environmental change accumulates. At some point, accumulated environmental change creates conditions under which the desired ecological condition likely collapses.

Promote resilience to change – Resilient ecosystems accommodate gradual changes related to climate, but they tend to return to a prior condition after disturbance. Ecosystem resiliency is increased by managing ecosystem attributes that help communities and populations cope with disturbance. For example, strategies that influence tree regeneration in forests, water table recharge in wetlands, or fire frequency in fire-adapted ecosystems will increase ecological resilience by resetting ecological conditions to a previous, more resilient ecological state. These strategies are best suited to ecosystems that have high amenity or commodity values, support endangered species, or are relatively insensitive to climate change effects. Maintaining ecological resilience to change may require more intensive interventions as changes in climatic conditions accumulate over time. Despite efforts to promote increased ecological resilience to change, these strategies have a high likelihood of failure over time. At some point, accumulated environmental change creates conditions under which these ecosystems may no longer rebound to their desired ecological state.

Enable ecosystems to respond to change – Manage ecosystems that enable communities and populations to respond to expected changes and result in new species composition under accumulated environmental change. For example, strategies that facilitate species and community dispersal or strategies that enhance local environmental heterogeneity may facilitate acceptable outcomes in response to changed conditions. In these cases, the goal is to encourage gradual and acceptable ecological transition to inevitable environmental change in order to avoid future ecological collapse. These strategies mimic, assist, or enable ongoing natural adaptive processes such as patch dynamics, species dispersal, metapopulation dynamics, changes in community composition, and disturbance regimes. Unlike the previous two approaches, which revolve around resistance to and recovery from change, this approach manages ecosystems so they respond to environmental change that is expected to result in new species combinations. This ecosystem response can move either toward some general direction (deterministic), where specific goals are planned for the future, or in unknown directions (indeterministic), where future goals are developed to accommodate uncertainty.

Realign management and restoration approaches to reflect future conditions – Design and implement management approaches to anticipate future changes in environmental conditions rather than to reflect current or past conditions that shaped ecological communities. This approach looks to future predicted climatic regimes, especially the predicted climatic surprises and threshold effects that can have extraordinary impact on community structure (Perring et al. 2013), and implements management strategies that ecologically pre-position communities for the future. Restoration that can anticipate future climatic impacts and design accordingly are not without controversy, especially those involving novel plant communities (Murcia et al. 2014).

These four approaches are hardly exclusive of one another, and they can play multiple roles at any given site. Still, site limitations and restoration goals may constrain restoration options available to increase future resilience, and restoration must be adapted accordingly. Perhaps the biggest obstacle that prevents incorporating climate change resilience strategies into restoration is the element of uncertainty (Staudinger et al. 2015). While there is little uncertainty about the inevitability of climate change, different models often provide different answers to questions regarding the magnitude and timing of potential changes. These uncertain and often conflicting details confuse us, the restoration community, when perhaps they should not. At minimum, restoration strategies can focus on the impacts that are certain. The climate is warming, creating longer growing seasons and more extreme weather events (Staudinger et al. 2015). Seasonal and spatial precipitation patterns will be impacted as well (Mishra et al. 2010). The interactions between these manifestations of climate change will influence the stressors that designers of restoration projects can plan around. Extreme precipitation events are likely to create new surface and hydrologic interactions (Pryor et al. 2014). Restoration sites can be designed to accommodate and hold excess water. In headwater streams, restoration can enhance the capacity of adjacent floodplains based on predicted storm intensities (Figure 23). If designed and implemented coherently across watersheds, these restorations can influence downstream reaches, wetlands, and lakes as well. These extreme storm events may be predicted to create new stream morphologies in otherwise natural habitats that can be addressed by reconfiguring or restoring adjacent floodplains. Wetlands can be designed specifically to accommodate anticipated severe storm events.

Droughts may increase in areas where summer precipitation decreases while average temperature increases, producing increased drought stress (Pryor et al. 2014). Wetland habitats for amphibians and reptiles can be designed to accommodate a variety of hydroperiods to provide insurance under intensive drought in the future. Wetland restorations can be designed to reestablish and enhance hydrologic gradients and ecotones across adjacent

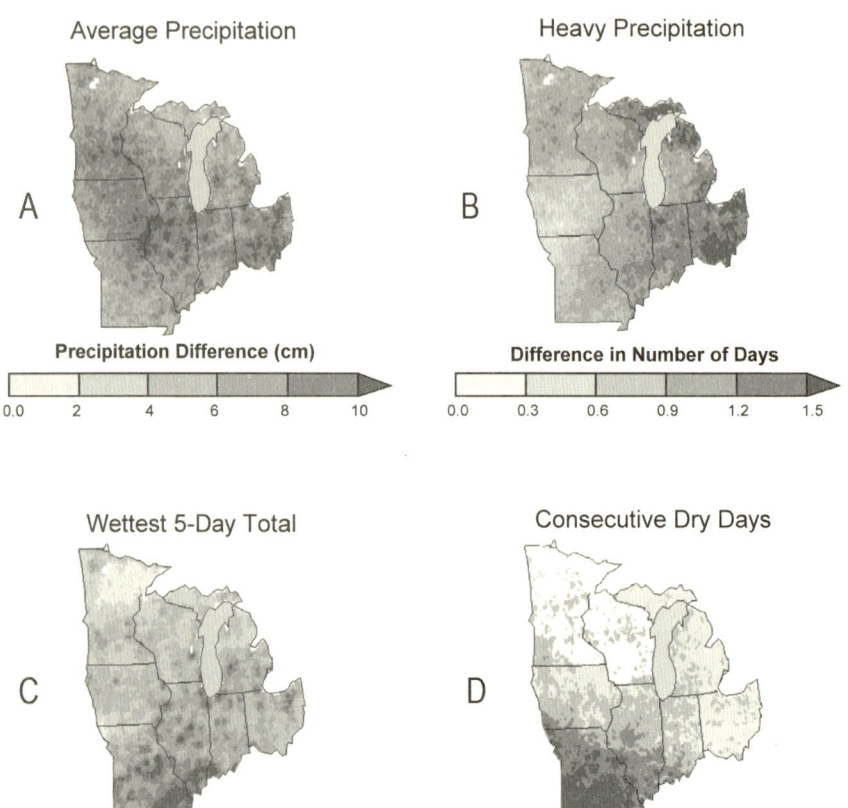

FIGURE 23. Predicted changes in precipitation in the Midwest under continued emissions. Differences in precipitation metrics were calculated as the difference of the time period from 2041 to 2070 and the time period from 1971 to 2000. The amount of precipitation from rainfall and snowfall is projected to increase (A). An increase in the number of days with very heavy precipitation (top 2 percent of all rainfalls each year) is also expected to occur (B). Increases in the amount of rain falling in the wettest five-day period over a year are predicted to occur (C). B and C indicate that heavy precipitation events will increase in intensity and frequency in the future across the Midwest. Change in the average maximum number of consecutive days each year with less than 0.03 centimeters of precipitation (D) indicates an increased potential for drought in the future. (Adapted from Pryor et al. 2014. Source: NOAA NCDC/CICS-NC.)

upland habitats to enhance resilience. Reestablished plant communities can incorporate species or genotypes that may be drought adapted. Existing plant communities can be managed to enhance drought-resistant aspects in anticipation of future climatic regimes. Restorations can reconnect isolated habitat fragments across local landscapes to increase the availability of microclimatic conditions that provide ecological refuges that buffer against future drought.

Severe storms with extreme winds or tornadoes may result in localized catastrophic damage to isolated terrestrial sites, especially forests (Saunders et al. 1991). In smaller forest fragments, habitat values may be severely impacted by such localized events simply because of forest size. Restoration can be used to increase forest size at such sites, creating habitats that are more appropriately scaled to absorb such rare but potentially catastrophic events. The consequence of more rainfall is felt differently across different land-use types. For example, in agricultural areas, much of the rainfall is transferred to streams via a network of subsurface drainage pipes. In these areas, which cover large areas of the Midwest, the largest flood peaks have not changed substantially but volumes have increased greatly (Lenhart et al. 2011). In contrast, the northern, forested part of the Midwest has not experienced substantial stream flow increase in recent decades.

Incorporating climate change into restoration designs requires geographically explicit predictions, as opposed to the simplistic generalization presented above. There are multiple models in use by climate change scientists, each with their own baffling sets of assumptions, algorithms, and nuances (Randall et al. 2007). It seems unfair that we, as designers of restoration projects, also must become experts in climate modeling to make the best decisions for our sites. Fortunately for us, these models all predict future climates based on similar inputs and are calibrated by the historical climatic record. As a result, they generally predict similar climatic futures. Thus, it is possible to use websites such as Climate Wizard (http://www.climatewizard .org/) to calculate "ensemble average" models that combine outputs from several techniques into a potentially more robust forecast (Araujo and New 2007). While no climate change model can perfectly predict the future, these consensus outputs can be used to identify climate change–induced stressors at an appropriate geographic scale for restoration planning.

The above thoughts and generalities provide an overview of the generalized impacts of climate change on a regional basis. However, ecological
restoration is a site-based practice that requires localized projections that
can be incorporated into specific informed decisions at the local level. The
Midwest is a climatically and ecologically diverse region, and typically,
restoration goals are shaped by local reference conditions, which, given
their development during past climatic regimes, intuitively function to
integrate across historic conditions (Howell at al. 2012). Our purpose is to
provide a conceptual framework that allows planners to accommodate the
impact of future climatic regimes as well. As designers and planners, our
professed goal should be to produce restorations that are a success today
and in the future. Incorporating climate adaptation strategies requires that
we also look to the future using the following stepwise process.

1. Use climate models to obtain the predicted future climates at the
   appropriate spatial scale for the project. The geographic precision
   of climate modeling and the unpredictable nature of weather itself
   require regional, not site-specific, examination of model outputs.

2. Translate the climate model predictions into ecological stressors
   that will impact ecological restoration goals under predicted climatic regimes in the future. The models should provide insights
   into future climatic deviations from current and past regimes,
   and how these may play out at the site level. In the Midwest, key
   factors typically revolve around increased temperatures during
   key growing or dormant seasons, hydrologic impacts from altered
   precipitation patterns, and the interactions between these climatic
   components (Pryor et al. 2014).

3. Prioritize which stressors will likely have the greatest potential
   impact at the restoration site relative to restoration goals. Dramatic events such as large rainfall events or severe drought are
   likely to have immediate impacts at sites. Additionally, gradual
   climate changes such as lengthening growing season and increased
   average temperature will have more subtle ecological impacts that
   accumulate over time. The interaction between the dramatic and

gradual changes will create the climate-associated stressors that impact ecological communities at the site level.

4. Develop and implement restoration design strategies that are likely to increase ecological resilience at the site level such as (a) engineered approaches to better accommodate severe hydrologic events; (b) increased ecological scale to accommodate catastrophic events; (c) increased connectivity and enhanced microhabitat diversity to buffer against long-term climatic changes; (d) genetic enhancement and novel species introductions in anticipation of future climates; and (e) transformational restoration and management that anticipates future conditions.

Not all sites may easily lend themselves to climate adaptation restoration strategies. Given that the goal of ecological restoration is to enhance ecosystem health, integrity, and sustainability (SER 2004), not increasing resilience to future climatic regimes whenever possible is ethically wrong. As such, restoration ecologists and practitioners have an obligation to consider ecological performance into the foreseeable future using the best available tools and science at their disposal.

APPLICATION OF GENERALIZED RESTORATION STRATEGIES AT REAL-WORLD RESTORATION SITES

In this chapter we present three examples that highlight restoration strategies for terrestrial ecosystems that address climate change at multiple spatial scales. Restoration strategies for aquatic ecosystems are addressed elsewhere in this volume (see Chapters 2, 4, 5, 6, and 9), and these strategies can be adapted to address climate impacts, as appropriate, using the simple guidelines above. In many respects, the examples we present are initial experiments in climate adaptation ecology, and only with the passing of time will we fully understand how these experiments faired in achieving their objectives. All were implemented as demonstration sites from which others can learn and observe in the hope that climate change adaptation strategies become more widely used in the future. It is important to remember that the climate-centric restoration design criteria and goals presented here address a subset of the larger restoration goals at each site. Site- and region-specific limitations and goals are likely to have a dominant influence on shaping restoration design and potential performance. However, given the range of design strategies and decisions available, including seemingly simple choices such as selecting appropriate seed ecotypes, it is hard to imagine restoration projects in which some level of climate adaptation could not be incorporated.

Northern Minnesota Forest Management – Shifting the Boreal Mosaic to Northern Hardwoods

Northern Minnesota's forests sit at the ecotone between the boreal and temperate forest zones and tallgrass prairie. Climate change is likely to produce profound changes across this iconic 20,000-square-kilometer landscape. At the time of European settlement, the region consisted of a complex mosaic of flat, poorly drained landforms dominated by peatland vegetation, including bogs, black spruce (*Picea mariana*) and tamarack (*Larix laricina*) swamps and fens, interspersed within glacially scoured uplands supporting red and eastern white pine (*Pinus resinosa* and *Pinus strobus*) forests and boreal forests of jack pine (*Pinus banksiana*) and black spruce (Galatowitsch et al. 2009). A legacy of ecologically incompatible logging and fire suppression has homogenized the region's forests, leaving them vulnerable to climate change and other emerging stressors. The structural and compositional diversity of the past has given way to second-growth stands dominated by aspen (*Populus tremuloides*), black spruce, and balsam fir (*Abies balsamea*) mixtures, which are species common in the boreal forests northward into Canada. Tree species that are seemingly adapted to thrive under future predicted warmer and drier climatic regimes (e.g., bur oak [*Quercus macrocarpa*], bitternut hickory [*Carya cordiformis*], black cherry [*Prunus serotina*]) are not well positioned to replace existing boreal forests because these forests are situated on the prairie-forest border. Changes in average annual temperature and precipitation by 2069 suggest a shift in regional climates equivalent to the current conditions found approximately 400 to 500 kilometers south-southwest of the project site (Galatowitsch et al. 2009). Predicted future climates for northern Minnesota forests will include warmer summers with more frequent and longer droughts. Average temperatures are projected to increase 3°C. Although annual precipitation is projected to increase slightly, a declining trend in summer precipitation is anticipated, resulting in increased drought stress (Galatowitsch et al. 2009; Pryor et al. 2014).

The future of northern forests is uncertain under these warmer, drier conditions. Boreal forests may be lost from Minnesota, while cold-temperate deciduous forests may persist only on north slopes in northern Minnesota. Species such as black spruce, white spruce (*Picea glauca*), balsam fir, tamarack, and paper birch (*Betula papyrifera*) are likely to experience severe declines (Landscape Change Research Group 2014). The rate of climate change will likely outpace the ability of well-adapted species such as bur oak, bitternut hickory, and black cherry to disperse and occupy newly available habitats (Scheller and Mladenoff 2008; Fisichelli et al. 2014). The northern forest ecosystem may well self-organize without appropriate tree species for future climatic regimes (Cornett and White 2013).

The Nature Conservancy, the Northern Institute of Applied Climate Science, and the University of Minnesota Duluth have developed and implemented a restoration strategy to increase the climate resilience of future forests to enable these ecosystems to continue to provide critical societal and economic services. The working forests of northern Minnesota underpin much of the regional economy, ranging from ecotourism (i.e., birding, hunting, fishing) to traditional forestry production. The general public recognizes the positive ecosystem services provided by the forested landscape and wants to maintain these ecosystem services into the future. Four strategies were used to enhance climate resilience: (1) ecologically based forest management practices that enhance ecological diversity at both the site and landscape level were implemented. These practices were designed to maintain and enhance existing species and structural diversity; (2) native tree species that would likely thrive under predicted future climatic regimes were identified and established in restorations across the region. These species are likely not able to disperse northward naturally due to landscape fragmentation and seed dispersal limitations without facilitated dispersal into restorations; (3) the genetic diversity of plantings was enhanced by introducing seedlings with genomes native to a neighboring region in the south and west. This ecological region is warmer and drier than the restoration area, and gene pools from this region are assumed to be preadapted to future climatic conditions; and (4) performance outcomes such as seedling survival, growth, and phenology were closely monitored to develop best practices for adaptation in northern forests (Cornett and White 2013).

Minnesota's northern forests will experience dramatic change over the next century. These proactive restorations in harvested units are intended to increase ecological resilience and to facilitate ecological response to change such that northern forest communities have the potential to self-reorganize gracefully into coherent, native botanical communities. By increasing species and structural diversity in existing forests, local plant communities are better positioned to respond to climate change impacts. Facilitating the dispersal of native species northward increases the potential reorganization pathways available to forested landscapes. Future forests will be different from those we know today, but the benefits they provide such as habitat for wildlife and wood for wood products should continue into the future.

Transitioning Mesic Forests to Drought-Resistant Oak Woodland Mosaics

Upland forests across much of the Central Hardwoods region, which includes the Ozark Mountains of Missouri and the unglaciated portions of Illinois and Indiana, are increasingly vulnerable to future climate impacts. This is especially true in light of recent changes that have occurred in the Central Hardwoods region over the last few decades, where decades of selective timber management, fire suppression, and herbivore mismanagement have produced forest communities that exhibit altered forest structure and tree composition that is more vulnerable to future predicted climate regimes (Brandt et al. 2014). These process manipulations have created and continue to create forests with a greater density of trees and plants associated with the cooler, damper, and more stable microenvironments of a closed canopy forest (in place of plants associated with the warmer and drier conditions of an open canopy forest) (Nowacki and Abrams 2008). These emerging forests are dominated by mesic, fire-intolerant species that are vulnerable to future predicted climatic regimes.

In south central Indiana, open woodlands dominated the Highland Rim and Shawnee Hills Natural Regions (Homoya and Huffman 1997, Homoya 1997). Droughty, bedrock-derived soils and frequent wildfires maintained an open woodland structure dominated by drought-tolerant and fire-resistant oaks (*Quercus* spp.) and hickories (*Carya* spp.). In localized areas, the woodland gave way to open graminoid barrens and glade communities, and the ecotonal interface between these community types was extensive.

In the mid-twentieth century, fire suppression created an ecological change that continues today. Woodlands that were once fire and drought adapted are replaced by maturing stands of trees with sparely vegetated forest floors. This change resulted from a positive feedback loop consisting of increased canopy cover that promoted increased soil moisture that led to fuel reductions that inhibit wildfire. Where drought-tolerant, open oak hickory woodlands with fuel-rich forest floor communities once thrived, closed canopy, moist beech (*Fagus* spp.)-maple (*Acer* spp.) forests are spreading upslope from floodplains and narrow valleys, slowly eliminating drought-tolerant communities that once dominated the well-drained hills and ridges. This change has been called the "mesophication" of eastern forests by many forest managers and scientists (Aldrich et al. 2005).

Although climate change will impact forest ecosystems in a variety of ways, it is likely that the most important direct impact of climate change will be drought stress (Gustafson and Sturtevant 2013), especially the occasional but long-lasting and severe drought events likely to occur under future predicted climate regimes (Dale et al. 2001). Virtually all modeled future climate scenarios predict that south central Indiana's climate will become warmer and wetter on an annual basis. Predicted average temperature increases range from 2° to 5°C by mid-century for Indiana and much of the Midwest. Similarly, climate change models predict an increase in annual precipitation ranging from 5 to 20 percent for much of Indiana. However, predicted precipitation will be highly seasonal, with dramatic increases in the dormant season and declines during the growing season relative to historic conditions. This inverse relationship between seasonal temperatures and precipitation will create increased summer drought stress. Higher growing season temperatures will increase evapotranspiration rates during the season that available soil moisture is predicted to be at its lowest. More importantly, these conditions will set the stage for potentially ruinous prolonged droughts that have the potential to severely impact forest ecosystems. Mishra et al. (2010) projected changes in the duration of drought periods in Indiana over the next century, and most of the climatic models projected an increase in drought duration and the spatial extent of droughts. These results indicate that future droughts may shift from localized conditions to more regional phenomena. If the process of mesophication continues unabated, then middle-aged mesophied forest communities will be vulnerable under future climate regimes because of the low adaptive capacity of their dominant species to extreme drought (Brandt et al. 2014). In the worst-case scenario, mesic forests across the region could collapse because there are few available seed sources of drought-tolerant species present on the landscape.

In response to this future threat, The Nature Conservancy initiated a long-term restoration program that is intended to influence forest composition and structure to manage for diverse communities that are better adapted to withstand future climatic conditions. While it sounds intuitively plausible, reintroducing fire alone cannot reverse decades of suppression and the resulting mesophication. In the preceding decades, mesic seedlings have matured into pole-sized trees that are reaching the canopy. While most problematic mesic trees are susceptible to prescribed fire as seedlings and saplings, at some point fire becomes less effective at thinning stands because large mature trees are less easily killed by fire. So, while forest managers could increase the frequency of prescribed fires to reverse decades of fire suppression, increasing the fire frequency may create an unhealthy approach to managing forests and could damage timber quality on production forestland.

The Nature Conservancy's structural restoration program was designed to increase ecological complexity at forested sites in the Highland Rim Natural Region (Homoya et al. 1985) of south central Indiana. On drought-prone habitats such as xeric ridgetops and south- or west-facing slopes, mesic tree species are mechanically thinned or removed from the canopy and midstory, creating an oak woodland habitat structure with 80 percent canopy closure. Mesic forest species growing in cooler ravines and on north-facing slopes were not disturbed. Following structural restoration, prescribed fire is used to manage for fire-tolerant tree recruitment, enhance woodland structure, and maintain light penetration to the forest floor to enhance graminoid cover and density. This combination of thinning and controlled fire increases graminoid fuel load in combination with fuel from increased dry oak leaf litter, creating a positive ecological feedback that can sustain fire-adapted, drought-tolerant oak-hickory woodlands. At this stage, periodic prescribed fire can be used to advance regeneration of the oak-hickory forest and inhibit mesophication. Solely thinning the forest canopy without the use of prescribed fire does not alter the forest regeneration toward drought-tolerant species (Figure 24). Prescribed burns help maintain over time the structure and composition of dry or mesic forests resilient to climate change.

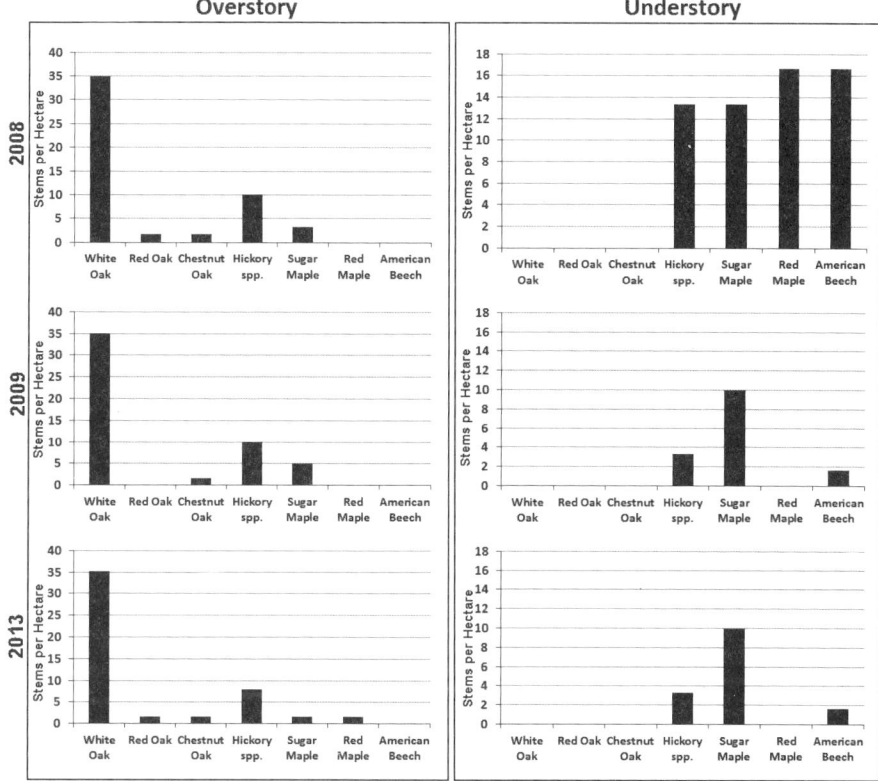

FIGURE 24. Response of drought-tolerant versus drought-intolerant tree species in the overstory and understory due to adaptation management in Brown County, Indiana. Prior to management in 2008, the overstory was dominated by drought-tolerant oak and hickory species while the understory was dominated by drought-intolerant maple and beech species. In 2009, mechanical treatment reduced but did not eliminate the recruitment of drought-intolerant species into the understory. In 2013, following prescribed fire treatments, drought-tolerant species began to become dominant in the understory. This trend is expected to continue as time passes, and it is anticipated that additional recruitment will occur in the next planned assessment, in 2018.

By working to adapt forests in southern Indiana to future climate conditions, The Nature Conservancy is taking advantage of the fact that these forests still have the components necessary to be resilient to climate change. We can create a complex mosaic of drought-adapted woodlands in drought-susceptible habitats while maintaining more mesic communities in more sheltered microhabitats. If future climates play out as currently anticipated, we believe that pre-positioning complex native communities on the landscape will allow dynamic ecological responses to play out (Maynard and Brewer 2013; Mawdsley et al. 2009) and will create different but high-quality native forested ecosystems in Indiana.

Indiana Sand Prairie Oak Barren Mosaics – Restoring the Hydrologic Gradient and Mosaic Structure

Savanna and associated prairie and wetlands were once common habitats across much of North America. These ecosystems consisted of complex habitat mosaics of open, emergent wetlands, sedge (family Cyperaceae) meadows, mesic sand prairies, xeric oak savanna ridges, and everything in between. These sand prairie and savanna mosaics are well known for supporting regionally and globally rare plants and animals, and they have been a high priority for conservation throughout much of the Midwest. For example, in Indiana The Nature Conservancy has aggressively pursued restoration and conservation at two major macrosites within the sand prairie and savannas in the region, Kankakee Sands and Tefft Savanna. At Kankakee Sands, The Nature Conservancy initiated a landscape-scale restoration to create an eighty-nine-square-kilometer contiguous conservation site that reconnects three scattered conservation areas (O'Leary and Shuey 2003; Shuey 2013). While climate change adaptation was fully incorporated into the Kankakee Sands restoration, here we describe a targeted series of projects adjacent to the Tefft Savanna Complex at the Indiana Department of Natural Resources' Jasper-Pulaski Fish and Wildlife Area that were initiated to increase ecological resilience of state-owned conservation lands.

In the early 1800s, the project area was predominantly black oak (*Quercus velutina*)–dominated barrens with fewer than fifty trees per hectare. Approximately 25 percent of the area was wet prairie and marsh, which occurred at elevations below 215 meters (Haney et al. 2008; Bowles et al. 2011). Today, most privately owned wet and mesic habitats are in agricultural production and contain a network of agricultural drainage ditches that lower near-surface water tables from adjacent low-lying sandy soils and adjacent sand rises. The upland dunes, which once supported open oak barren communities, now support a tree density in excess of 400 stems per hectare, the apparent result of an extended period of fire suppression (Haney et al. 2008). The sand prairie mosaic of northwestern Indiana is often thought of as a relictual community that developed a few thousand years ago under conditions perhaps more extreme than projected climate regimes (Transeau 1935) and maintained initially by aboriginal fire. As such, this ecosystem mosaic is presumed to be preadapted to future climates that are likely to resemble the Hypsithermal period.

Consensus climate modeling using Climate Wizard yielded predictions of warmer, wet winters and warmer, drier summers in northwestern Indiana. Periodic drought stress will likely become a major ecosystem driver in the region. The most critical impacts to wildlife include drought mortality in plant communities and seasonal loss of wetland habitats, especially for amphibians (Broadman et al. 2006). Climate change models predict that multiyear events will override annual fluctuations in shaping wildlife communities in the region (Brandt et al. 2014).

Based on those threats, over the last decade The Nature Conservancy acquired a series of tracts totaling approximately 3.3 square kilometers known collectively as the Prairie Border Nature Preserve. These agricultural tracts were purchased so that restoration efforts could target agricultural drainage systems that lower the near-surface water table within the adjacent Jasper-Pulaski Fish and Wildlife Area and impact wetland hydrology (Figure 25). There are three distinct components to this restoration effort. The agricultural drainage ditches reduce the hydrologic storage of wetlands within the adjacent Jasper-Pulaski Fish and Wildlife Area. By eliminating the drainage network, the ecological resilience of adjacent lands to future predicted climatic regimes can be enhanced by restoring the natural hydrologic regime to these adjacent areas. This was accomplished by simply filling the existing drainage ditches using the original dredge materials that still lay alongside them.

The agricultural fields were restored to enhance ecological resilience as well. This is accomplished by cessation of agricultural production and creating subtle topographic depressions in the sandy soils to form seasonally inundated wetlands on these former agricultural fields. In these

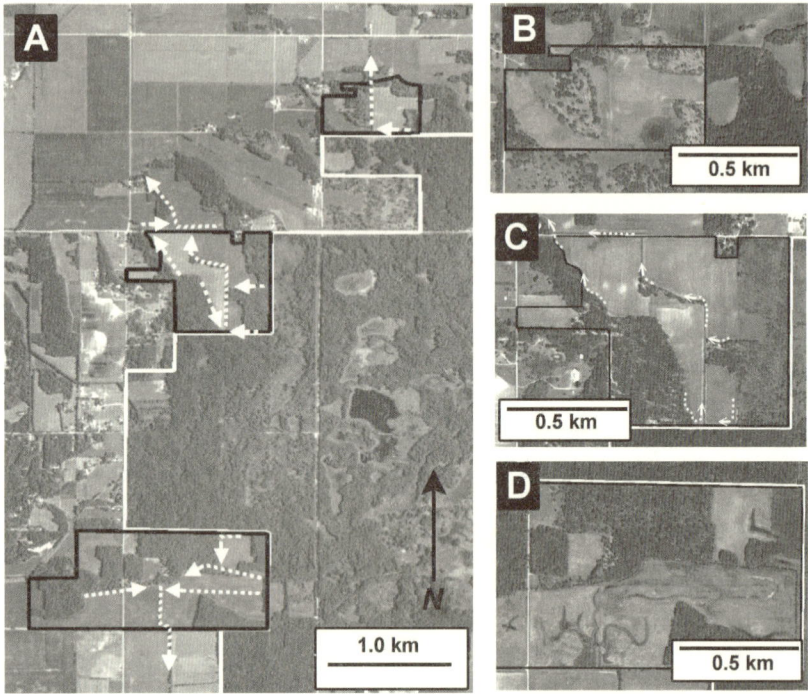

FIGURE 25. Hydrologic and structural restorations implemented adjacent to the Jasper-Pulaski Fish and Wildlife Area. The eastern conservation area (outlined in white) indicates the tracts identified as having hydrologic impacts on adjacent conservation lands (outlined in black) (A). The white arrows represent agricultural drainage ditches impacting adjacent lands (A). The northernmost tract immediately following hydrologic and structural restoration in winter 2014–2015 (B): note the ditch elimination and immediate wetland response in the southeastern corner of tract as well as extensive structural restoration of upland barrens (B). The central tract immediately following structural restoration during winter 2014–2015 and prior to hydrologic restoration scheduled for autumn 2015 (C). The primary agricultural ditch is clearly visible in the center of the tract (C). The southernmost tract following hydrologic restoration in 2004 and moderate structural restoration (D).

settings, the near surface water table seasonally rises and falls, and typically inundates depressions from the dormant season through midsummer. By varying the depth of these depressions, we created a variety of hydroperiods, including a few wetland pools that are likely to persist during periods of prolonged drought.

Zones within the hydrologic gradient were seeded into appropriate high-diversity plant communities in order to initiate the creation of a dynamic habitat mosaic. Three primary local genotype seed mixes were developed for wet, mesic and dry sand communities using over 150 native species to establish an initial community mosaic based on nearby habitat remnants (i.e., reference ecosystem concept; Chapter 2). We used plugs to establish late successional wetland communities such as sedge meadows and bare-root seedlings to establish pin oak (*Quercus palustris*) along wetland margins, and black oak in dry upland plantings. We established species-rich, high-diversity restorations in an attempt to create enhanced ecological redundancy and climatic tolerance within and across community types.

These agricultural tracts also include expanses of mesophied oak barrens on sand ridges that were too xeric for production agriculture. Mechanical clearing of the canopy to approximately 60 percent cover was used to kick-start structural restoration, followed by three years of herbicide treatment of resprouting trees and shrubs. By increasing light penetration to the ground, we released the existing herbaceous seed bank and promoted the emergence of the sparse survivors of the open savanna herbaceous community that lay dormant under shaded conditions. The herbaceous community responded quickly to increased light levels and fire management.

After initial restoration and establishment of the herbaceous community, these tracts are managed with prescribed fire to enhance herbaceous communities and to open the canopy in the black oak barrens further. The goal of this restoration effort is to increase ecological resilience by restoring the ecological complexity and dynamism across the hydrologic gradient, both within the restoration sites themselves and the adjacent conservation lands that were impacted by agricultural drainage practices. By reestablishing natural hydrologic dynamics combined with fire management, we expect to create a habitat mosaic that can respond dynamically to future climatic conditions. We expect that high-quality plant communities will be able to "slide" up or down hydrologic gradients as they adjust to future climatic regimes.

CONCLUSIONS AND SUMMARY

The three examples discussed highlight strategies that seemingly increase ecological resilience to climate change, but they are really intended to set the stage for ecosystems to respond gracefully to climate change. No one expects that our terrestrial plant communities of today will emerge from the next century unchanged. Nor do we expect to somehow lock these communities into their current state by some miracle of restoration and management. Ecological communities are dynamic and are expected to respond as such to impending and ongoing changes to climate. Some ecosystems, such as the boreal forests of northern Minnesota, are likely to experience profound change. Superficially, the forest of the twenty-second century may have little in common with those that we know today. Our job as designers and planners of restoration projects is to ensure that the difference is indeed superficial. The oak woodlands and black oak savannas of Indiana are seemingly less vulnerable to dramatic change. In reality, they are not, and the strategies described above are intended to facilitate a dynamic response.

Our ecosystems of the future must support biologically rich native plant communities, provide habitat for native wildlife, and support local economies through tourism, outdoor recreation, and timber products. Additionally, these ecosystems must continue to provide ecosystem services such as high water and air quality to contribute to quality of life for regional residents. In northern Minnesota, this will likely require moving southern species that are unable to disperse northward due to land-use fragmentation. In Indiana oak ecosystems, it will require an initial reshuffling and enhancement of existing ecosystem components. Designers and planners of ecological restoration projects may think of themselves as stacking the ecological deck toward success (Cornett and White 2013)—at least for the first round of play. Creating the conditions that will allow for resilient, self-organizing ecosystems is important for ensuring that restored ecosystems have the ability to adapt to near-term climate impacts.

None of the examples discussed long-term monitoring or assessment, as the critical role of adaptive management is implicit. The ecological impacts of climate change will unfold in ways we cannot fully comprehend now. Restoration-based climate change strategies are at best a professional guess of the response of ecological communities in the face of uncertain climatic projections (Lawler 2009). Our actions are a first play in a long game that will span professional careers. Too many restoration projects are based on inappropriate timetables that assume that ecosystem development is controllable and predictable (Hilderbrand et al. 2005). This is especially true of restoration projects performed as mitigation with time-bound performance criteria. Such projects routinely fail as native habitats after initial performance criteria are met (Race and Fonseca 1996) and ecological management lapses. Climate impacts will only exacerbate this problem. Restoration efforts that succeed will do so because we are watching and reacting.

Restoration as a climate change adaptation strategy diverges from traditional ecological restoration. Traditionally, practitioners have defined restoration goals and outcomes based on historical community structure and function (Howell et al. 2012) and success was based on attainment of those properties. Goals for climate adaptation are based on future conditions that may diverge wildly from the conditions that shaped current or historical communities (Harris et al. 2006; Choi et al. 2008). Restoration success will be based upon the ability to adapt to those conditions as they emerge. We cannot predict exactly how climate change will impact our sites. Our restorations, designed for resilience and adaptive capacity, will become experiments in adaptive management or adaptive restoration (Zedler 2000b). Success demands that we protect our restoration investments with monitoring and intervention if ecological trajectories are not proceeding appropriately (Hilderbrand et al. 2005; Choi et al. 2008). Our responsibility as designers and planners is to lay the groundwork of increased resilience to climate change. Our ultimate success will depend upon our successors continuing the restoration process using our initial strategies as their entrée to the challenge ahead.

CHAPTER EIGHT

INVASIVE SPECIES MANAGEMENT

Developing a Common Vision for Midwestern Landscapes

DAN SHAW

INTRODUCTION

Invasive species including plants, animals, microbes, and pathogens are causing significant challenges for restoration professionals across the Midwest (see Chapters 1–7). They are organisms that are not native to the ecosystem being restored, and their introduction can cause economic or environmental harm or harm to human health (National Invasive Species Council 2001; National Invasive Species Council 2008). With over 50,000 individual invasive species documented in the United States, we are witnessing pronounced changes to the structure and function of native plant communities. These changes are raising new questions about how to most effectively prioritize invasive species control and focus restoration and available resources to maintain biodiversity and important ecosystem functions and services. The Midwest has a strong legacy of conservation that provides a foundation for addressing these new challenges, but innovative approaches are needed to address specific invasive species threats. Collaborative efforts are playing a key role in prioritizing invasive species, sharing key information, finding common ground for management, and engaging the public. My objective is to review the legacy of invasive species management in the Midwest and propose key strategies for finding a common vision for invasive species prioritization and control.

Between 10,000 and 20,000 years ago, the last period of glaciation made its retreat north, leaving the foundation for the Midwest's lakes, rivers, and wetlands, as well as a diversity of natural habitats. A unique and varied assemblage of plant communities with characteristics unique to the Midwest developed over time within the prairie, deciduous forest, and coniferous forest biomes of the region. Examples of these plant communities include conifer swamps, peatlands, floodplain forests, marshes, wet meadows, deciduous forests, savannas, and prairies. The location and composition of these plant communities were influenced by unique climate, natural disturbance patterns, geology, soils, hydrology, and plant and animal interactions. These plant communities were also home to a diverse array of wildlife species (Tester and Keirstead 1995).

European settlement in the Midwest brought rapid changes to the landscape that drastically reduced ecosystem functions and services, and created favorable conditions for invasive species. These disturbances included the drainage and alteration of wetlands, near elimination of bison and other wildlife, plowing of prairies, fire suppression, and extensive logging. In the case of tallgrass prairie, less than 0.001 percent of the original land cover remains (Kurtz 1996). Ohio, Indiana, Illinois, Iowa, and Missouri have all lost more than 85 percent of their wetland area (Dahl and Allord 1997). These changes, along with decreasing wildlife populations and degrading air and water quality, led to the work of early conservationists such as John Muir, Gifford Pinchot, Hugh Bennett, Sigurd Olson, Aldo Leopold, and Rachel Carson, which would significantly influence the Midwest. Aldo Leopold introduced the term *land ethic* in his book *A Sand County Almanac* (Leopold 1949), in which he focused attention on the role that people play in caring for the land. His emphasis on the importance of biodiversity and intact ecosystems is well represented in his quote "A thing is right when it tends to preserve the integrity, stability, and beauty of the biotic community. It is wrong when it tends otherwise" (Leopold 1949).

Early ecology research and natural resource conservation work created a foundation for the field of ecological restoration (see Chapters 1 and 2), but we still have many challenges to address as our landscapes continue to change in ways that favor invasive plants, animals, and pathogens. These challenges include landscape fragmentation, lack of natural disturbances (fire, flooding, grazing, etc.), altered community structure, nutrient inputs, altered hydrology, and soil disturbance. Climate change is introducing a new set of variables (warming temperatures and increased storm intensity) that influence landscape stability (see Chapter 7). Together, these changes are leading to the development of altered ecosystems exhibiting unique combinations of plant species. These ecosystems are a source of debate about our potential for success in restoring highly altered ecosystems and managing them to maintain diversity and important ecosystem functions and services. The potential to reclaim degraded ecosystems is dependent on the degree of alteration, available resources, and the motivation of people who value them.

WHY SO MANY INVASIVE SPECIES?

New invasive species continue to arrive and create challenges for already overburdened natural resource managers. How are species getting here? Can the pathways leading to their arrival be minimized or eliminated? Individual species have been intentionally and randomly introduced to the Midwest. Pathways for unintended introductions include release of ballast water from ships, aquaculture escapes, boats and other recreational vehicles, ornamental plants that escaped from gardens, contaminated seed, waterway connections, soil movement, roadway transportation, and escaped domesticated animals such as feral pigs (*Sus scrofa*).

Canada thistle (*Cirsium arvense*) may have been one of the earliest unintended releases (Table 3), having arrived in contaminated seed in the 1600s and impacting agriculture and grazing (Sheley and Petroff 1999). Brown rats (*Rattus norvegicus*) originating from northern China were transported to Europe around the mid-1500s and to North America around 1755 via the shipping industry (Lack et al. 2013). They have led to significant ecological and financial impacts in many countries where they have established. More recently, the expansion of international trade has had unintended consequences for midwestern forests with the introduction of harmful nonnative insects. The emerald ash borer (*Agrilus planipennis*) and Asian long-horned beetle (*Anoplophora glabripennis*) currently eating their way through midwestern forests are believed to have been introduced from wood packing materials from Asia (Herms and McCullough 2013).

TABLE 3. Time line of selected invasive species' arrival to the United States.

Date of Arrival	Name	Origin and Means of Spread
1600s	Canada thistle (*Cirsium arvense*)	Origin: Europe/Asia Contaminated grain seed, hay, and ship's ballast (Sheley and Petroff 1999)
Mid-1700s	Brown rat (*Rattus norvegicus*)	Origin: Europe/Asia Accidental introduction as stowaways on ships (Lack et al. 2013)
1840s	Common carp (*Cyprinus carpio*)	Origin: Asia Intentional release to enhance wild fish populations (Moyle 1986)
1850s	House sparrow (*Passer domesticus*)	Origin: Europe/Asia Intentional release by directors of the Brooklyn Institute (Williams 2002)
Mid-1800s	Common buckthorn (*Rhamnus cathartica*)	Origin: Europe/Asia Intentional introduction as an ornamental shrub (Broennimann et al. 2014)
1860s	Burning bush (*Euonymus alatus*)	Origin: Asia Intentional introduction as an ornamental shrub (Flinn et al. 2014)
1868 or 1869	Gypsy moth (*Lymantria dispar*)	Origin: Europe/Asia Intentional release to develop a silk industry (Liebhold et al. 1992)
1880	Common starling (*Sturmus vulgaris*)	Origin: Europe/Asia/North Africa Intentionally released as part of an effort to introduce all of the species mentioned in Shakespeare's works in the New World (Tibbetts 1997)
1875	Japanese barberry (*Berberis thunbergii*)	Origin: Asia Introduced as an ornamental shrub (Lavoie et al. 2005)
1884	Spotted knapweed (*Centaurea stoebe*)	Origin: Europe/Asia Accidental introduction in contaminated seed (Liebhold et al. 1992)
1940s	Kudzu (*Pueraria lobata*)	Origin: Europe/Asia Introduced as an ornamental ground cover and erosion control (Tibbetts 1997)
2002	Emerald ash borer (*Agrilus planipennis*)	Origin: China Accidental introduction from infested wood packing materials (Herms and McCullough 2013)

Several shrubs, including Asian bush honeysuckles (*Lonicera* spp.), common buckthorn (*Rhamnus cathartica*), glossy buckthorn (*Frangula alnus*), and Amur maple (*Acer ginnala*), were intentionally introduced as landscaping plants and for soil stabilization (Broennimann et al. 2014). It is estimated that 85 percent of woody invasive species in North America were introduced for the landscape trade (Reichard and Hamilton 1997; Snow 2002). Other species that now dominate large areas of the Midwest and decrease floral diversity and pollinator resources include reed canary grass (*Phalaris ardundinacea*) and smooth brome grass (*Bromus inermis*). These two species were planted widely to stabilize roadsides and grass waterways, and for grazing and haying. Common carp (*Cyprinus carpio*) was originally introduced into midwestern waterbodies as a potential food source (Moyle 1986), and it now reduces sportfishing opportunities by destroying spawning and reproduction of walleye (*Sander vitreus*) and other species such as largemouth bass (*Micropterus salmoides*) and northern pike (*Esox lucius*). Gypsy moth (*Lymantria dispar*) was intentionally released in the United States with the intent of it interbreeding with silkworms (*Bombyx mori*) to develop a silk industry (Liebhold et al. 1992).

DEVELOPING A COMMON RESPONSE

Discussions about how to prioritize species and how to respond to new invasions are constantly evolving. Agricultural weed control likely started soon after European settlement in the eastern United States and focused on weeds that decreased crop yields. Overgrazing of rangelands led to degradation of plant communities, erosion, and the introduction of several invasive species including Canada thistle, spotted knapweed (*Centaurea stoebe*), and cheatgrass (*Bromus tectorum*). The Taylor Grazing Act of 1934 led to the creation of grazing districts and the use of managed grazing, and these steps were later followed by efforts to restore rangelands (Ross 1984). The first attempts to control invasive animals in the United States focused on species such as rock pigeons (*Columba livia*) and brown rats that were established in the Midwest by the 1700s and damaged crops (Lack et al. 2013).

Invasive species control for nonagricultural reasons is a more recent phenomenon, likely beginning with the management of gypsy moths in

Massachusetts (Liebhold et al. 1992). This insect was present for twenty years before the population spiked, causing widespread damage to urban and rural trees. This population increase led to the development of a commission to oversee efforts to exterminate the species through chemical treatment and natural enemies. These early efforts were not successful and caused severe damage to native silk moth populations. They subsequently were replaced with the use of gypsy moth egg mass removal and trapping (Dunlop 1980; Liebhold et al. 1992).

Early prairie restoration efforts that started with Curtis Prairie at the University of Wisconsin–Madison Arboretum around 1930 also involved invasive species control (see Chapter 3). Portions of Curtis Prairie had been used as a horse pasture and had a high percentage of Kentucky bluegrass (*Poa pratensis*). The dominance of invasive species within Curtis Prairie created challenges during early vegetation establishment efforts. Despite the success of this early restoration effort, this site continues to be influenced by invasive species through the impacts of reed canary grass and leafy spurge (*Euphorbia esula*) on floral diversity levels.

Research conducted over the last several decades has played a key role in understanding how invasive species change plant community structure and natural disturbance patterns, and how these changes influence our ability to reach restoration goals. The effects of invasive species can be measured in biodiversity and basic ecosystem functions such as nutrient cycling, soil development, and energy transfer. Changes to these ecosystem functions influence the ability of ecosystems to provide critical ecosystem services such as water purification, support for rare and endangered species, timber production, fisheries health, and recreation (see Chapter 9). For example, nonnative earthworms (suborder Lumbricina) gradually consume the organic content of soils in maple (*Acer* spp.)-basswood (*Tilia* spp.) forests, changing nutrient cycles and leading to the loss of tree seedlings, wildflowers, and ferns (class Polypodiopsida) (Frelich and Holdsworth 2002). In Midwest wetlands, nonnative common reed (*Phragmites australis*) has formed dense stands that impede the flow of water and accumulate sediment. Zebra mussels (*Dreissena polymorpha*) displace native freshwater mussels (family Unionidae, family Sphaeriidae) and interrupt food webs by decreasing phytoplankton and zooplankton populations. Nonnative trees and shrubs colonize opportunistically and host few native insects, subsequently providing only minimal food sources for native songbirds

(Tallamy 2004). Maintaining and restoring ecosystem function and services is central to the practice of ecological restoration, which makes invasive species control essential to the success of future restoration efforts. The influence of invasive species also goes beyond the field of ecological restoration, as they also impact agriculture, forestry, grazing, and recreation. Maintaining ecosystem services is a common goal for each of these natural resource disciplines and serves as a starting point for identifying a shared vision for invasive species control.

Recent invasive species efforts have focused on assessing the risk of invasive species, preventing their introduction, and initiating control of new species after their initial detection. Collaboration between scientists and practitioners has been essential for understanding which invasive species may cause the greatest change to the physical structure and function of ecosystems and which require a greater priority for control. As more native plant communities have been restored over the last several decades and more invasive species have arrived, restoration professionals have become selective in their invasive species control efforts to get the greatest benefit for the invested time and funding. Some invasive species are tolerated due to a lower risk to ecosystem functions, while other species that represent a greater risk form the focus of invasive species control efforts. To aid decision-making, individual states have been developing risk assessments to evaluate the likelihood of introduction and establishment, extent of existing populations, dispersal potential, and possible harmful effects (Buerger et al. 2016). To complement these efforts, researchers have been developing predictive models. For example, ecological niche modeling was conducted in Iowa to help environmental managers assess the risk of invasive species due to climate change (Ingenloff et al. 2011). Predictive models were developed and tested to assess invasion potential for six species: round goby (*Neogobius melanostromus*), red swamp crayfish (*Procambarus clarkii*), Asian rock pool mosquito (*Aedes japonicus*), parrot-feather (*Myriophyllum aquaticum*), Chinese bushclover (*Lespedeza cuneata*), and New Zealand mudsnail (*Potamopyrgus antipodarum*). These species were selected due to their history of colonization in regions outside of their native habitat, their potential threat to Iowa's biodiversity, and available data. The risk potential for each species was projected for three emissions scenarios to the years 2050 and 2090. The species considered to have a high invasion risk were the red swamp crayfish and the Asian rock pool mosquito (Ingenloff et al. 2011).

Once invasive species priorities are identified, the next step is to develop effective control and management programs. The National Invasive Species Management Plan developed by the National Invasive Species Council in 2001 and updated in 2008, as well as the USDA Forest Service Invasive Species Program, have helped partnerships across the United States find a common vision by identifying key components of successful invasive species control programs. Key components include (1) prevention, (2) early detection and rapid response, (3) control and management, (4) restoration, and (5) organizational collaboration. The development of a framework for invasive species control has helped professionals across the Midwest coordinate their efforts. Below is a summary of the main aspects of each of these key components of effective invasive species control programs.

Prevention

Preventing the invasion of priority species is the best method of control, as the majority of invasive species are nearly impossible to eradicate once they become established. Most states have developed invasive species prevention programs, which represent a good investment to reduce future costs. Outreach is a critical part of prevention and ensures that landowners, nonprofit organizations, and industry are vital partners in keeping invasive species out. The PlayCleanGo awareness campaign, with over 300 partners across the United States, is a good example of effective outreach. A variety of outreach methods are used to communicate key messages to a diverse audience, including billboards, radio, TV ads, booths at fairs, and landowner workshops. These key messages include the removal of plants, animals, and mud from boots, gear, pets, and vehicles; cleaning gear before entering and leaving recreational sites; staying on designated roads and trails; and using certified local firewood and hay.

Prevention efforts may also include the development of state and federal noxious weed lists and local ordinances that require agencies to enforce these policies. Several midwestern states evaluate the likelihood of introduction and establishment, extent of existing populations, dispersal potential, and possible harmful effects resulting from invasive species (Buerger et al. 2016). Additional prevention efforts include the development of best management practices to guide wetland restoration, forestry practices, and work within transportation corridors.

Early Detection and Rapid Response

Effective prevention programs can make a significant difference, but they cannot stop the spread of all invasive species. Early detection and rapid response are the next line of defense. These efforts focus on eradicating or containing invasive species before they become widespread. Invasive species may exhibit population trends like the gypsy moth's, which maintained a stable population and then exponentially increased (MISAC 2015). Predicting which species may become invasive is difficult. Japanese barberry (*Berberis thunbergii*) and burning bush (*Euonymus alatus*) were sold by nurseries for years before their populations significantly increased in the Midwest. Tracking populations of insects, fungi, and small aquatic organisms is more difficult than tracking plants, and new methods are needed to help determine when these more mobile types of invasive species pose a risk. A combination of research, risk modeling, and monitoring is needed to understand the characteristics of invasive plants, animals, and pathogens and to predict how much of a threat their establishment represents for terrestrial and aquatic ecosystems. Citizen scientists are playing an increasingly important role in helping to monitor and report invasive species before they become a bigger problem (Crall et al. 2010). They add to the workforce needed for these early detection efforts to be effective. New online tools are becoming available to help with reporting invasive species too. For example, EDDMapS is a collaboration among multiple stakeholders working to rapidly respond to new invasive species in midwestern states. The website, along with a smartphone and tablet app, facilitates efforts to report invasive species locations. The development of rapid response teams is also an important part of early detection efforts. For example, the Southern Illinois Invasive Plant Strike Team is funded by state and federal sources and works full time to respond to early detection events and control invasive species in pathways of spread before they can threaten natural areas.

Control and Management

Invasive species control and management include rapid response efforts and sustained long-term efforts to manage established species. Eradication and containment are reasonable goals if species are not well established. Attempts to contain a species involve frequent monitoring, management of area boundaries, and occasionally restrictions on recreational use in the containment area. Invasive species control efforts for more established populations focus on locations that have been identified as a high priority based on risks to valued natural resources. There is often a focus on restoring the most intact areas first and then working outward to control the species. This approach helps protect areas that are providing important ecological functions and services.

Control and management methods can be categorized as (1) cultural practices, (2) physical restraints, (3) removal, (4) judicious use of chemicals and biopesticides, (5) release of selective biological control agents, and (6) interference with reproduction (National Invasive Species Council 2001). Multiple control methods may be combined as part of management efforts. The Midwest Invasive Plant Network has developed an Invasive Plant Control Database summarizing control information collected through scientific review, expert opinion, and user input.

Use of control methods along with prevention is often referred to as integrated pest management. Integrated pest management focuses on increasing effectiveness of invasive species control while protecting natural resources. A good example of integrated pest management is the recent effort to control common carp. Common carp were introduced to the Midwest in the mid-1800s (Table 3) and are now widespread in lakes and rivers. This fish species causes severe impacts to shallow lakes and wetlands by disrupting shallowly rooted aquatic plants and releasing phosphorus. These ecosystem changes negatively impact native fish and waterfowl populations and communities (see Chapter 4). Traditional control methods for common carp have involved nonspecific toxins and water level drawdowns that may impact non-target species in wetland and stream ecosystems. Recent efforts to reduce common carp populations in the Midwest focus on combining the practices of suppressing reproduction by introducing game fish, radio-tagging to track movement, controlling movement into spawning areas, and strategically planned harvests (Colvin et al. 2012).

Restoration

Prevention, early detection, rapid response, and control efforts all help maintain the integrity of ecosystems in the Midwest. When ecosystems become dominated by invasive species, it can become difficult to restore key ecosystem functions and services. Recent scientific findings indicate that community structure (i.e., biodiversity) may be essential for the maintenance of stable ecosystem productivity (Tilman et al. 2006; Biondini 2007).

Ecosystem restoration increases plant community resiliency to invasive species by increasing diversity and overall ecosystem health. Increasing ecosystem resiliency is a relatively new focus for restoration. These goals have been part of restoration efforts for many years, but they are being applied in new ways such as favoring different combinations of species or varying the time and frequency of disturbance. An example would be wetland restoration projects coordinated by the Minnesota Board of Water and Soil Resources in which restoration seed mixes, plant selection, and restoration methods are all planned to maximize landscape resiliency. Decision-making is based on a variety of factors including project goals, presence of invasive species, available native seed bank, expected hydrologic conditions, and potential for long-term resiliency to invasive species. Project assessments, monitoring data, and discussion between conservation partners play an important role in developing restoration plans and adapting restoration techniques over time (Figure 26).

Organizational Collaboration

There are many benefits of collaboration and sharing information between international, federal, state, local, and tribal governments, private organizations, and individuals. Federal policy positions for invasive species can be communicated to international organizations as part of agreements such as the Convention on Biological Diversity, International Plant Protection Convention, and environmental cooperation mechanisms developed through free trade agreements. Through federal and state work groups, agency budgets can be compared to maximize the use of funds and request additional funding as needed. Collaboration on federal, state, and local levels is also key to identifying grant opportunities and research needs. Many partners are also working together on outreach and education campaigns to maximize resources and reach a wider audience. Partnerships are also playing a key role in setting invasive species priorities and increasing the effectiveness of control efforts in a wide range of landscapes across the Midwest (National Invasive Species Council 2008).

FIGURE 26. Discussions between natural resource professionals play a key role in finding common ground for invasive species management.

FUTURE CHALLENGES

What else might assist invasive species getting a foothold in the Midwest? The future is uncertain, and many professionals are working to understand how potential changes to climate and cultural practices will influence the integrity of terrestrial and aquatic ecosystems across the Midwest. Future challenges for invasive species control consist of climate change, an expanding world economy, landscape fragmentation, and changes to how people view and interact with our natural landscapes.

We do not fully know how climate change will influence ecosystems, but warming temperatures have led to invasive plants, animals, and pathogens expanding their range distributions northward. Species that have often been considered invasive species typical of the southeastern United States, such as kudzu and water hyacinth, have invaded the Midwest. Nonnative grasses (family Poaceae) used as biofuels may pose a greater invasive species risk with an expanding range (Jørgensen 2011). Pine bark beetles (family Scolytidae) that have led to large-scale losses of pines (*Pinus* spp.) in western states now threaten pine forests across the Midwest. Invasive insects may also develop earlier in the growing season, allowing them to produce more generations each season. An extended growing season benefits invasive plant species like common buckthorn that are active early and lose their leaves later than native shrubs. Warming air temperatures and flashier rain events may cause the loss of native species within plant communities and leave gaps that can be occupied by invasive plants having greater dispersal and colonization abilities than native plants. An increase in extreme storms favors species such as common reed, narrowleaf (*Typha angustifolia*) and hybrid cattails (*Typha × glauca*), and reed canary grass, which benefit from increased sedimentation and nutrients. These invasive aquatic plant species handle water fluctuations better than many sedges and forbs, and their establishment leads to decreased plant diversity (Green and Galatowitsch 2002).

Many invasive species originated in Europe and Asia (Table 3) and lack natural parasites and predators that limit their population levels. A recent study suggests that invasive species from some parts of the world such as East Asia have become superior competitors in their new environments because they have been evolving over millions of years with no recent periods of glaciation (Fridley and Sax 2014). Global trade invariably introduces new species from around the world. Between 1985 and 2000, nonnative bark beetles were intercepted at U.S. ports from imported cargo on 6,825 occasions by the USDA Animal and Plant Health Inspection Service (Lee et al. 2007). Ballast water from European and Asian shipping has already brought over 140 species to the Great Lakes (Ricciardi and MacIsaac 2000). Packing materials are another potential source of invasive species that have led to the recent introduction of emerald ash borer (Table 3).

Continued urban development and agricultural expansion will likely lead to additional fragmentation of intact ecosystems, decreasing their resiliency and making them more vulnerable to invasive species. The increased edge of fragmented ecosystems provides more opportunities for the introduction of invasive plants, animals, and pathogens. Fragmentation problems are compounded when accompanied by impervious surfaces and subsurface drainage that increase the transport of nutrients (nitrogen, phosphorus) and promote the establishment of invasive plant species such as reed canary grass and common reed.

Additionally, our connections with natural landscapes have been changing across the Midwest due to busy lifestyles and less time spent outdoors. It is becoming more common that landowners do not live on the land they own for agricultural or hunting purposes. Many landowners may only set foot on their property a few times each year. Across the Midwest, small farms with farmers who have an intimate knowledge of their land and a strong relationship with local conservation staff have become less common. Changes in landownership affect the feeling of personal responsibility toward an individual piece of property and can influence the amount of effort that will be put toward invasive species management in the future. As invasive species spread across property boundaries, it is important to have engaged landowners with a personal stake in control efforts.

DEVELOPING A FUTURE VISION

Future challenges will require strong partnerships and coordinated efforts to find solutions and a common vision for how to control invasive species. Federal, state, and local policies can facilitate early detection and rapid response efforts. The public also plays a crucial role in supporting invasive species control. As our landscapes adjust to development, climate change, and other factors, we will need to make difficult decisions about where to allocate limited resources. We will also need to adapt to changing conditions to meet new challenges and increase plant community resiliency to invasive species. Forming strong multisector partnerships, advancing invasive species policy, and coordinating invasive species planning, public engagement, and adaptive restoration planning will all play important roles in future years.

Forming Strong Multisector Partnerships

Strong partnerships between federal, state, and local agencies, conservation groups, landowners, and industry partners are becoming more common. Cooperative weed management areas (CWMAs) are diverse partnerships that come together to manage invasive species across geographic areas. They help identify priorities, spread out the workload, and guide effective prevention, early detection, and rapid response efforts. There are over sixty

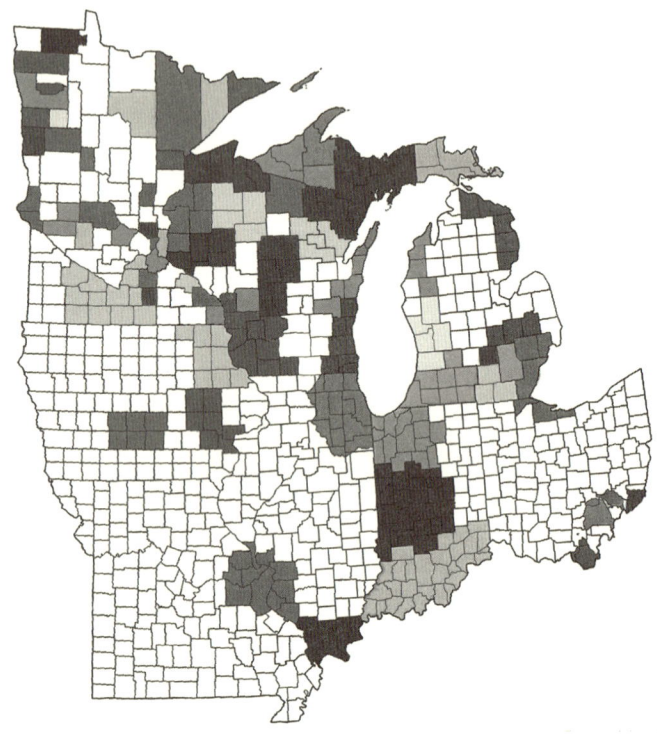

FIGURE 27. Location of weed management organizations across the Midwest with different shades of gray showing where one group transitions to another. (Adapted with permission from Midwest Invasive Plant Network.)

local weed management organizations across the Midwest, many of which are part of CWMAs (Figure 27). These weed management areas vary from one to several counties in size and develop plans to prioritize target species and identify areas for control. The Midwest Invasive Plant Network has played a key role in assisting the development of midwestern CWMAs and other local weed management organizations by providing resources and training opportunities.

The Becker County CWMA in Minnesota is an example of a strong partnership that works to find common ground for invasive species management. It was started in 2006 as a multiagency and community effort. Located in northwestern Minnesota, the Becker County CWMA consists of federal, state, tribal, the Nature Conservancy, county, and private properties, and these properties contain some of the most imperiled prairie habitats in the United States. Partners include the Becker County Soil and Water Conservation District (program lead), townships, watershed districts, state and federal agencies, The Nature Conservancy, agricultural chemical companies, city and county staff, the University of Minnesota Extension, White Earth Indian Reservation, and a local all-terrain vehicle club.

The Becker County CWMA partners developed a plan to define objectives and key invasive species control strategies that included planning and management, education, inventory and mapping, early detection and rapid response, and use of an integrated pest management approach. Mapping invasive plant populations has been critical to developing a comprehensive invasive species management plan. An early detection system coordinated by multiple partners identifies species that have the potential to become established within the Becker County CWMA. The Becker County CWMA is committed to integrated pest management including mechanical, chemical, biological, and cultural methods of control. For example, a combination of chemical, biocontrol, and replanting with native species is used to control spotted knapweed. As new threats, control opportunities, and priorities emerge, the Becker County CWMA modifies its plan. An example is the revision of planning and roadside management practices to address the rapid movement of wild parsnip (*Pastinaca sativa*) within the county. Annual work plans are developed to guide efforts each year. The CWMA has played a key role in protecting and restoring intact prairies and other habitats in the county by preventing the spread of invasive plants and participating in invasive species control efforts (Henry and Watland 2009).

In addition to the broad approach to invasive species management adopted by CWMAs across the Midwest, partnerships have formed to address specific invasive species challenges. The Great Lakes Phragmites Collaborative formed to coordinate the management of common reed, a highly invasive plant species that is now common in North American wetlands. The collaborative has focused on coordinating communication among a wide range of stakeholders: private landowners, lake associations, road commissions, state and federal agencies, and private conservation organizations. Their advisory committee includes representation from federal, state, and provincial governments and others engaged in invasive common reed management, research, restoration, education, and outreach throughout the Great Lakes Basin. As part of its outreach efforts, this partnership has developed a website that provides detailed information on biology, ecology, control, identification, monitoring, mapping, and recent research related to common reed. Webinars, blog posts, and a listserv are also used as part of the outreach efforts to provide information about specific restoration projects where successes and failures can be shared to aid other initiatives across the region.

Advancing Invasive Species Policy

State and federal legislation and local ordinances need to address invasive species challenges because they can effectively promote early detection and rapid response efforts. Ordinances focus attention on the importance of controlling invasive species and coordinating efforts between partners, leading to progressive action and effective early detection and rapid response.

Currently there is not one federal law or program that provides a list of problematic invasive species for regulation. So invasive species programs tend to delegate listing authority to a state agency. As a result, individual midwestern states including Illinois, Indiana, Michigan, Minnesota, Missouri, Ohio, Wisconsin, and Iowa have noxious weed laws and have developed their own lists of species that pose risks to agriculture, public health, and the environment (Environmental Law Institute 2004). Formally listing invasive species in combination with noxious weed laws and other regulatory requirements is an effective prevention method because it increases awareness of the problem. These lists need to be updated and consistently enforced. Once on the list, the invasive species then trigger requirements that can be enforced at multiple levels of government (Environmental Law Institute 2004). States have been setting up detailed assessment methods to determine the potential risk of individual invasive species and their potential for listing. Regulations then require control of species or restrict their sale and transport.

The Wisconsin Department of Natural Resources was directed by the legislature in 2001 to establish a statewide program to control invasive species. The Wisconsin Invasive Species Council was formed in 2004. By 2009, the Wisconsin Department of Natural Resources created Wisconsin's Invasive Species Identification, Classification, and Control Rule. This rule guides the public to identify and minimize the spread of invasive plants, animals, and diseases that can damage Wisconsin's land and waters. Lists were developed for fungi, algae, cyanobacteria, plant disease–causing organisms, plants, and terrestrial and aquatic animals. Wisconsin allocated its Department of Natural Resources more authority to provide a statewide law that coordinates invasive species control efforts. A science-based method is being used for classifying species into the categories of prohibited or restricted. With few exceptions, the transport, possession, transfer, and introduction of prohibited species is banned. For restricted species, there is a ban on transport, transfer, and introduction, but possession is allowed with the exception of fish and crayfish. Regulations are aimed at preventing new invasive species from entering Wisconsin and enabling quick action to control or eradicate newly arrived ones.

Coordinating Invasive Species Planning

Strong invasive species planning efforts tend to involve multiple partners and a coordinated effort to solve specific challenges. Partnerships formed for invasive species control need to set well-defined goals. Containment of invasive species is likely a realistic goal, but total eradication may not be possible in many cases. Low levels of some invasive species can be tolerated if they do not significantly alter ecosystem structure and function. A key step in the planning process is to assess available resources (i.e., staff, equipment, funding) and determine how to maximize these resources to meet project goals. It is common to prioritize ecosystems with the greatest need for management. For example, restoring a habitat corridor or complex can be one of the best strategies for increasing landscape resiliency to climate change and invasive species while protecting imperiled plants and animals.

Multiple agencies and organizations have joined forces to address large threats to ecosystem integrity of midwestern forests caused by several invasive insects. Gypsy moth is a destructive species that was intentionally introduced into the United States in 1868 or 1869 (Table 3). It is currently established throughout the Northeast and parts of the upper Midwest, where it feeds on over 300 tree species, with oaks (*Quercus* spp.) being a preferred species. One hundred and twenty million square kilometers of midwestern forests have been defoliated by gypsy moth since 1970. Congress funded the Slow the Spread Program in 2000. In cooperation with the USDA Forest Service, ten states (Ohio, Indiana, Michigan, Wisconsin, Minnesota, Indiana, North Carolina, Virginia, West Virginia, and Kentucky) located along the leading edge of gypsy moth populations have implemented a regionwide strategy to minimize the rate at which gypsy moth spreads into uninfested areas. As a direct result of this program, the spread of gypsy moth has been dramatically reduced by more than 70 percent from the historical level of twenty-one kilometers per year to around five kilometers per year. To meet its goals, the program conducts intensive monitoring with pheromone-baited traps that can detect isolated or low-level populations of gypsy moths. The program also places an emphasis on the use of control methods to suppress gypsy moth populations (i.e., removal of infested trees, insecticide treatment, girdling ash trees, use of environmentally benign tactics such as increasing natural predator abundance) (Leuschner et al. 1996).

Emerald ash borer is an invasive insect species from Asia that arrived in southeastern Michigan in 2002 and has killed an estimated thirty million ash trees (*Fraxinus* spp.) there. Subsequently, this insect species has spread to fourteen other states and parts of Canada. The Slow Ash Mortality Program was piloted in Michigan and is another example of an effort to control a damaging forest pest, with an emphasis on monitoring. This program involves a collaboration between Michigan State University and federal, state, and local partners. The goal of the Slow Ash Mortality pilot project in Michigan's Upper Peninsula is to slow the local invasion process and allow land managers time to be proactive rather than simply reacting to overwhelming numbers of dead trees. Extensive surveys are conducted using girdled trees and traps to determine the extent of emerald ash borer populations. A variety of emerald ash borer population suppression tactics are used that include removal of infested trees, use of insecticides, and girdling ash trees. Continued research and the development of new control methods will create more options for emerald ash borer management and increase the effectiveness of existing technologies. Although many have concluded that emerald ash borer cannot be eradicated, slowing its growth can provide time to develop long-term responses. Research and field trials are under way to understand how emerald ash borer influences the amount of native and invasive species cover, light levels, and soil properties, as well as the types of tree species that can be planted as replacements to help revegetate forests and prevent invasive plant species from colonizing forest gaps (McCullough and Mercader 2012).

Public Engagement

Finding a common vision for invasive species control also involves engaging the public. Beyond just increasing awareness of the priorities that have been set for invasive species, effective campaigns communicate how individual members of the general public can get involved in invasive species prevention and control efforts. Some of the most effective outreach efforts are hands-on and actively involve the public in invasive species control. This approach makes the public aware of the importance of the issue and gives them the knowledge and tools to conduct invasive species control efforts in other areas. Ultimately, the public will play a key role in the future of invasive species control. When the public is informed and engaged, they can take action on private properties, push for local ordinances, support state and federal legislation related to effective prevention and control, and encourage the protection and management of local natural areas.

There are many good examples of effective public outreach efforts being led by organizations such as conservation districts, watershed districts, universities, agencies, and nonprofits. One nonprofit that has worked to engage the public since 1995 is Great River Greening, based in Saint Paul, Minnesota. The organization focuses on helping communities restore, manage, and learn about their natural environment through volunteer involvement across the Twin Cities region and other parts of Minnesota. Great River Greening trains volunteer supervisors who help lead multiple large volunteer restoration events each year. They have involved 25,000 volunteers over the past twenty years in events ranging from common buckthorn and garlic mustard (*Alliaria petiolata*) removal to replanting native trees and shrubs. Participants in Great River Greening volunteer events come away with an understanding of the importance of urban natural areas and an awareness of invasive species issues. Great River Greening has expanded its community engagement efforts to include a program for at-risk children, bringing them to outdoor classrooms to conduct hands-on restoration work and to teach them about job opportunities in the environmental field.

Other midwestern organizations are focused on conducting direct outreach to landowners to guide invasive species control. Conservation districts have played a key role in connecting landowners with conservation programs and funding across the Midwest since their formation in response to the dust bowl of the 1930s. There are over 400 conservation districts across the Midwest, and many of them are directly involved in CWMAs or have specific programs to address invasive species. For example, Wabasha Soil and Water Conservation District in Minnesota has played a key role in obtaining federal, state, and local funding to address the threat of Oriental bittersweet (*Celastrus orbiculatus*), an emerging weed threat. Conservation district staff are familiar with local conditions and develop strong relationships with local landowners. As a result, they can effectively communicate the importance of invasive species control and other practices that maintain ecosystem functions and services. Conservation districts also work closely in partnership with a wide range of local organizations and age groups. Collectively, they can have a positive influence on the future of ecosystem stewardship on private lands.

Adaptive Restoration Planning

Through years of field experience, restoration professionals have learned that being flexible and responsive to changing conditions is essential to the success of restoration efforts. Invasive species, habitat fragmentation, increased nutrients, pollutants, and climate change cause ecosystem stress and influence the success of restoration projects. Adaptive restoration planning is simply the use of an adaptive management approach in the design of an ecological restoration project. It begins with developing detailed restoration plans and schedules, and updating these plans as needed to adapt to future ecosystem responses and unexpected environmental challenges. In some cases, adaptive restoration planning to manage invasive species involves the use of innovations and new tools such as prescribed grazing, biocontrol agents, and novel combinations of restoration techniques. Monitoring is a critical component of any adaptive restoration plan. Like the financial audit of a fiscal plan, monitoring helps managers spend resources wisely and catches small problems before they escape and become big ones.

There are a wide variety of large-scale restoration efforts across the Midwest that have involved adaptive restoration planning and invasive species issues. In the prairie pothole region, partners from multiple states are focusing on science-based monitoring of restoration projects to battle invasive species such as Kentucky bluegrass, smooth brome grass, and reed canary grass that overtake prairies and wetlands and threaten ecosystem health by limiting the number of wildlife species such as grasslands birds, native bees (clade Anthophila), and butterflies (clade Rhopalocera) within these ecosystems. A multipartner team is leading an effort called the Native Prairie Adaptive Management Project to combine the expertise of scientists and land managers from multiple agencies and organizations. This coordinated effort is supported by the Plains and Prairie Potholes Landscape Conservation Cooperative, U.S. Geological Survey, and U.S. Fish and Wildlife Service. As part of their adaptive restoration plan, the Native Prairie Adaptive Management Project annually collects data from nearly 115,000 locations about the composition of native and nonnative vegetation and management actions from national wildlife refuges and wetland management districts in the prairie pothole region, which includes North Dakota, South Dakota, Minnesota, and Montana. This data is stored in a centralized database and can be used in predictive models to generate management recommendations to guide land managers on the timing of practices such as prescribed burning and grazing. The model will become more accurate as data is added over time, and eventually it will take some of the subjectivity out of decision-making (Gannon et al. 2013).

CONCLUSIONS

Aldo Leopold (1949) once wrote, "One of the penalties of an ecological education is that one lives alone in a world of wounds." It is clear that we are experiencing an exponential increase in invasive species populations that are impacting native plant communities across the Midwest. These changes are creating significant challenges for natural resource professionals, but we have come a long way in our approach to prioritizing and managing invasive species. Early efforts often arose out of necessity in order to protect agricultural crops and food stores. As the number of established invasive species increased and led to drastic changes within terrestrial and aquatic ecosystems, we began focusing more attention on maintaining ecosystem functions and services that are important for ecosystem health and our quality of life. Through our collaborative efforts, we have also become more unified in our approach to prioritization, prevention, early detection, control, and restoration. This common vision has increased our effectiveness and gives us much to be optimistic about as we look to the future.

CHAPTER NINE

THE EMERGING ROLE OF ECOLOGICAL RESTORATION IN AGRICULTURAL WATERSHED MANAGEMENT IN THE MIDWEST

CHRISTIAN LENHART AND PETER C. SMILEY JR.

Agricultural watershed restoration is the process of assisting the recovery of ecosystem structure and function within watersheds that have been altered and degraded by agriculture. Restoration of agricultural watersheds in the Midwest is challenging, in part because most of the land is privately owned. Many parts of the Midwest have 80 to 90 percent agricultural land cover, including northern Ohio, southern Michigan, the majority of Indiana and Illinois, as well as southern Minnesota and Wisconsin. These areas are mostly in row-crop agriculture with corn and soybeans and small amounts of small grains, hay, and cash crops. The Midwest also possesses the greatest concentration of agricultural drainage in the United States, with an estimated 10 to 60 percent of cropland containing subsurface drainage, agricultural drainage ditches, or both (Blann et al. 2009). Agricultural watersheds in the Midwest have experienced severe physical, chemical, and biological modifications as a result of land-use conversion, extensive stream channelization, and subsurface drainage for agriculture. As a result of this widespread and intensive agricultural land use across the region, there are numerous agricultural watersheds in need of restoration. Impairment of ecosystem functions in small headwater streams contributes to degradation and loss of aquatic habitats in larger rivers downstream via elevated loadings of sediments, nutrients, and pesticides and altered hydrologic regimes.

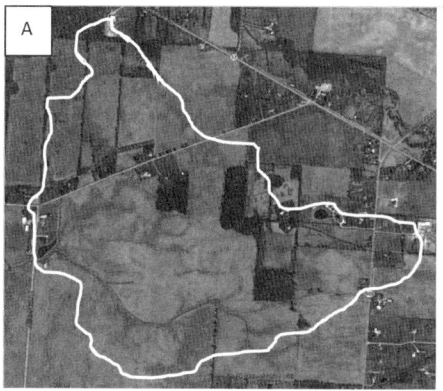 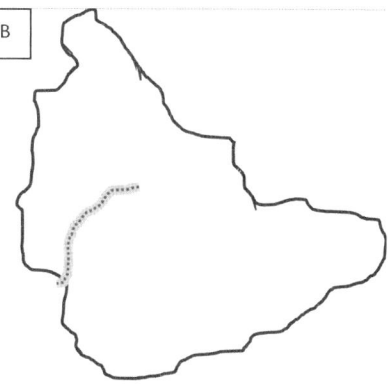

FIGURE 28. Aerial photograph (A) of a typical midwestern agricultural watershed with watershed boundary (white line) and a schematic of the same watershed (B) with upland habitat (light gray area), riparian corridor (dark gray area), and stream (dashed line) depicted.

Large-scale mechanization of agriculture discourages small patches of natural cover, windbreaks, and other seminatural areas (Leopold 1949), and as a result remnant habitats are scarce in the Midwest. In many parts of the Midwest, more than 98 percent of prairie and forests have been replaced with cropland, particularly in the southern half of the region (Blann et al. 2009). Additionally, within agricultural watersheds, cropland typically occurs in the uplands and the remnant habitats tend to be confined to narrow riparian corridors adjacent to the streams (Figure 28). The remnant prairies, wetlands, forests, and riparian corridors that remain are important for future restoration efforts.

Many site-level stream restoration projects have been implemented within degraded watersheds in the Midwest (Moerke and Lamberti 2004; Alexander and Allan 2006). Regionwide estimates on the cost of stream restoration are not available, but within Ohio, Michigan, and Wisconsin alone an estimated $444 million was spent between 1990 and 2004 (Alexander and Allan 2006). Common restoration goals for these projects included instream habitat improvement, bank stabilization, and water quality improvement (Moerke and Lamberti 2004; Alexander and Allan 2006). Despite these efforts, restoration at the watershed scale is the rare exception in agricultural regions of the Midwest due to production demands and an existing

TABLE 4. Types of ecosystem services (Millennium Ecosystem Assessment 2003) provided by agricultural watersheds and where the services originate within the watersheds.

Type	Ecosystem Service	Watershed Source
Provisioning	Source of food and fiber	Agricultural fields within the uplands
Provisioning	Source of energy	Agricultural fields within the uplands can be a source of biofuel, and forested habitats in uplands and riparian corridors can provide firewood.
Provisioning	Source of drinking water	Surface and groundwater in the uplands and riparian corridors
Provisioning/ Cultural	Fish and wildlife habitat. Fishing and hunting are a way of obtaining food and a recreational activity.	Within remnant habitats (prairies, forests, wetlands, etc.) in the uplands, riparian corridors, and streams and rivers
Regulating	Storage of water within the watershed to prevent flooding	Lakes, wetlands, soils, and groundwater within the uplands and riparian corridors
Regulating	Maintenance of water quality through water filtration	Wetlands and lakes within uplands and riparian corridors of streams and rivers
Regulating/ Supporting	Cycling and storage of carbon. Carbon storage is also considered a regulating service because of its role in regulating greenhouse gas emissions.	Prairie habitat, grassland habitat, and wetlands in the uplands; wetlands in the riparian corridor
Cultural	Real-world laboratories for teaching children and adults about nature, ecology, and agriculture	Uplands, riparian corridors, and streams and rivers
Cultural	Recreational areas for birding, canoeing, hiking, and other outdoor activities	Within remnant habitats (prairies, forests, wetlands, etc.) in the uplands, riparian corridors, and streams and rivers
Supporting	Nutrient uptake and transformation	Within wetlands and other riparian habitats and within streams and rivers

continued on next page

Supporting	Uptake, transformation, and dissipation of pesticides, pharmaceuticals, and other agricultural contaminants	Within wetlands and other riparian habitats and within streams and rivers
Supporting	Oxygen production	Forested habitats in the uplands and riparian corridors are likely to result in the greatest net production of oxygen.
Supporting/ Cultural	Biodiversity. Biodiversity is typically expressed as the number of species in an area, but it includes a wide range of measurements that describe the variety of life across all levels of biological organization, from the organismal level to the ecosystem level (Daily 1997). Biodiversity is also a cultural service in part due to its contribution to recreational activities.	Within remnant habitats (prairies, forests, wetlands, etc.) in the uplands, riparian corridors, and streams and rivers

cultural barrier resulting from the incorrect assumption that restoration will require the elimination of agriculture from these watersheds. It is also likely that many are not aware that watershed-scale restoration is a way of increasing the diversity of ecosystem services provided by agricultural watersheds, particularly those provided by non-crop areas. Notably, nine of thirteen ecosystem services (source of freshwater, fish and wildlife habitat, water quality maintenance, carbon cycling and storage, recreational areas, nutrient uptake, agricultural contaminant uptake, oxygen production, and biodiversity) originate primarily from non-crop habitats within agricultural watersheds (Table 4). In contrast, upland cropland in the Midwest primarily provides one ecosystem service (food and fiber) (Table 4).

Opportunities for agricultural watershed restoration exist despite the barriers. Federal agricultural conservation programs (e.g., Wetlands Reserve Program, Conservation Reserve Program, Wildlife Habitat Incentive

Program) provide landowners and farmers with cost-share funds for implementation of conservation practices that could contribute to agricultural watershed restoration. These conservation programs are voluntary, and consequently implementation efforts are not coordinated among programs or within a watershed to create a comprehensive watershed restoration plan. Additionally, the current program structure does not support the adoption of multiple practices because farmers and landowners are not allowed to receive funds for multiple practices at the same time. Balancing the need to maintain farmland productivity with the cost of implementing conservation practices as well as bureaucratic hurdles all contribute to farmers' reluctance to enroll in conservation programs. Despite these imperfections, conservation programs represent viable funding sources that can assist watershed restoration efforts.

There is a great deal of history and experience in midwestern watershed-based restoration in the Midwest to guide current watershed restoration projects. The early watershed conservation projects beginning in 1933 with the Coon Creek Watershed Project in Wisconsin mark the start of the research and practice of watershed management that continues to be conducted by state agencies, federal agencies, academic institutions, nonprofit organizations, and for-profit organizations in the Midwest. Not to be overlooked, the agricultural community and agencies (i.e., farmers, county soil and water conservation districts, drainage commissions, Natural Resources Conservation Service (NRCS) offices) have extensive experience in land and watershed management in support of agricultural production that can contribute to watershed restoration efforts. The agricultural community values land stewardship in addition to economics. This appreciation of land creates a common link between the agricultural community and other, more environmentally minded stakeholders that can form the basis of effective watershed partnerships.

Our objectives are to (1) examine the similarities between ecological restoration, ecosystem services, and traditional watershed management approaches; (2) review theoretical frameworks applicable for designing future watershed restoration strategies in the Midwest; and (3) highlight how the most suitable theoretical framework (i.e., the Treatment Train Approach) can be used to design future restoration strategies for midwestern agricultural watersheds. We conclude with a discussion of the benefits and limitations of the Treatment Train Approach and our concluding thoughts.

ECOLOGICAL RESTORATION, ECOSYSTEM SERVICES MANAGEMENT, AND TRADITIONAL WATERSHED MANAGEMENT: A COMPARISON

The approach to agricultural watershed management in the Midwest has changed over time since the early 1960s and 1970s, when farmers and drainage districts were actively channelizing streams without regard to the environmental impacts. Currently, there is a greater focus on the removal of sediment, nutrients, and pesticides from targeted watersheds to improve the water quality of downstream waterbodies. Ecological restoration, ecosystem services management, and traditional watershed management represent three approaches that can be applied to agricultural watershed restoration in the Midwest. While the overall goals of these three approaches are complementary because they seek to improve ecosystem health, water quality, and ensure the sustainability of agricultural watersheds, their specific goals and approaches differ considerably.

Ecological restoration has traditionally focused on the reestablishment of historical native plant communities or ecosystems to revive the biotic diversity and plants and animals that have become rare in degraded ecosystems. Traditionally, restoration was planned with the goal of reestablishing ecological structure for its own sake. The most recent definition of ecological restoration used by the Society for Ecological Restoration (SER 2004) incorporates a broader viewpoint and defines it as the process of assisting the recovery of an ecosystem that has been degraded, damaged, or destroyed. The ultimate goal is to promote the recovery of ecosystem structure and function (SER 2004). This approach has not been widely adopted within agricultural watersheds.

The ecosystem services approach focuses on reestablishment of ecosystem services and functions within agricultural watersheds (Table 4). This approach requires more attention to the details of how ecosystems function than watershed management does. However, it also focuses less on the recovery of ecosystem structure than ecological restoration. This emerging approach is becoming more widely promoted for use within agricultural watersheds.

Traditional watershed management focuses on water quantity and quality using hydrology and soil science principles as well as engineering approaches and practices developed by soil and water conservation districts and the NRCS. Hydrologic criteria such as reduction of peak discharge or total discharge are often goals of watershed management, with the assumption that water quality and biotic improvements in the downstream waterbodies will follow. This is the most widely used approach within agricultural watersheds in the Midwest.

Assuming the spatial scale of implementation is the same, the three approaches differ in their relative potential to improve degraded ecosystems' habitat quality, ecosystem function, and fidelity (i.e., similarity in physical structure and function to historical or reference ecosystems) (Figure 29). Ecological restoration has the greatest potential for improving habitat quality for the biota, reestablishing ecosystem function, and attaining fidelity because of its focus on the recovery of ecosystem structure and function with the use of historical conditions or high-quality reference ecosystems as the restoration target. In some cases, ecological restoration may be necessary to achieve desired ecosystem services such as nutrient removal from wetlands. For example, the loss of native vegetation cover may make restored wetlands leak more nutrients downstream. Traditional watershed management has the least potential to improve habitat quality, ecosystem function, and fidelity because of its more narrow focus on water quantity and quality, although the approach is evolving. The ecosystem services approach has a greater potential for improving habitat quality and attaining fidelity than traditional watershed management does because it is more likely to focus on increased biodiversity and provision of ecosystem services than traditional watershed management.

However, in the real world the typical spatial scale of implementation differs among the three approaches. Traditional watershed management focuses on improving downstream conditions, and it is usually applied at larger spatial scales than the other approaches (greater than one square kilometer). Ecological restoration is typically applied at small site scales (0.01 to one square kilometer) within agricultural watersheds because its focus is more intensively on the reestablishement of ecosystem structure and function. The spatial scale of the ecosystem services approach depends on the number and types of ecosystem services addressed. It could range widely, from small sites to the large watershed scale. The spatial scales of implementation among the approaches differ due to the level of effort and funds needed. Ecological restoration approaches require the greatest effort and funds per unit area. In contrast, traditional watershed management typically requires the least effort and has more dispersed use of funds, unless

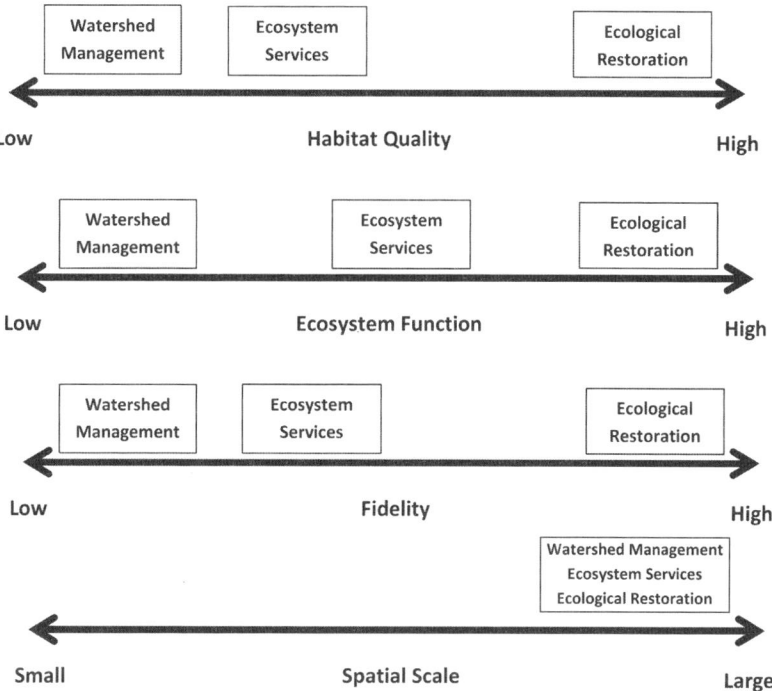

FIGURE 29. Relative potential of traditional watershed management, ecosystem services, and ecological restoration approaches to provide habitat quality, ecosystem function, and fidelity within agricultural watersheds when implemented at the same spatial scale.

water infrastructure (e.g., sediment control basins, levees, etc.) are used. The ecosystem services approach requires intermediate levels of effort and funds.

The three approaches may use many of the same practices. For example, practices effective at reducing surface runoff would be used in all three approaches if the objective is to reduce peak flows. A critical difference among the three approaches is their emphasis on providing benefits to people. Ecological restoration is more ecocentric than the other approaches because its goals are not centered on providing utilitarian benefits for humans (Jordan and Lubick 2011). Ecological restoration is the broadest and most holistic of the three approaches in the sense that it focuses on the recovery of ecosystem structure and function. The ecosystem services and traditional watershed management approaches have a more narrow focus than ecological restoration does, but they are more readily adopted within agricultural watersheds because of their greater emphasis on increasing the value of these watersheds for people.

The recovery of ecosystem structure and function represents the ideal for ecological restoration. Recent thinking in the field of ecological restoration has shifted its focus from re-creating the past to focusing on the future, with greater emphasis on recovery of ecosystem function and acceptance for novel structural conditions (Clewell 2009). It has been suggested that an insistence on re-creating past conditions may hinder ecosystem recovery (Apfelbaum and Haney 2010). Thus, within our definition of agricultural watershed restoration we emphasize it is the process of assisting the recovery of ecosystem structure *and/or* function within degraded agricultural watersheds. The use of this broader definition results in the ecosystem services and traditional watershed management approaches becoming alternative watershed restoration strategies because of their focus on improving ecosystem function or services. Our broader definition makes it clear that ecological restoration, ecosystem services management, and traditional watershed management are three separate approaches in a continuum of restoration strategies, which should assist with developing a wider range of watershed restoration strategies. Ideally, components of all three approaches would be used in watershed-scale restoration and management plans.

A REVIEW OF THEORETICAL FRAMEWORKS FOR DESIGNING WATERSHED RESTORATION STRATEGIES

We have identified five theoretical frameworks for designing restoration strategies for Midwest agricultural watersheds (Table 5). Four of these frameworks have either been implemented within agricultural watersheds in the Midwest (Naturalization, Riparian Management System), implemented in the Midwest (Treatment Train Approach), or developed for agricultural watersheds with extensive agricultural drainage (Building-Block Model). The fifth framework (Rosgen Natural Channel Design) constitutes one of the most widely used approaches to stream restoration design (Lave 2009).

Agricultural watershed restoration requires a watershed-scale approach to restoration design. None of the five frameworks explicitly incorporates

TABLE 5. Comparison of five frameworks for designing agricultural watershed restoration projects.

Approach	Description	Does approach explicitly promote management of the entire watersheds?	Does approach recommend specific practices?	What types of practices?
Treatment Train Approach (Broughton and Apfelbaum 1999)	Use of habitat creation and restoration throughout an urban development or watershed to reduce hydrologic transport time and facilitate uptake of contaminants	No	Yes	Upland, riparian corridor, stream
Naturalization (Rhoads and Herricks 1996)	An approach that sets restoration targets in light of human usage of streams instead of historical conditions; the goal is to develop restoration designs that establish sustainable, morphologically, and hydraulically varied yet dynamically stable fluvial systems that support biologically diverse aquatic ecosystems.	No	No	N/A
Building-Block Model (Petersen et al. 1992; Vought and Lacoursiere 2010)	A modular approach for restoring key functions and ecosystem services of agricultural streams that must continue to provide agricultural drainage; the approach involves five key practices: riparian restoration, subsurface drainage pipe alteration, in-channel modifications, creation of riparian wetlands, and daylighting of buried streams.	No	Yes	Riparian corridor, stream

continued on next page

Approach	Description	Does approach explicitly promote management of the entire watersheds?	Does approach recommend specific practices?	What types of practices?
Rosgen Natural Channel Design (Rosgen 1996, 2007)	Involves the use of a stream classification system based on geomorphology to assist with designing restoration projects; the focus is on designing stable stream channels following a forty-step procedure that is divided into eight phases: develop objectives, assess geomorphology and hydraulic conditions, conduct watershed assessment, consider changes in land-use management, design channels, select instream habitat structures, implement design, monitor and maintain.	No	Yes	Upland, riparian corridor, stream
Riparian Management System (Isenhart et al. 1997)	Uses a riparian management system consisting of four components (creation of multispecies riparian buffers, use of bioengineering stream bank stabilization methods, construction of wetlands designed to treat subsurface drainage runoff, and rotational grazing) that can be modified to fit site-specific conditions and landowner objectives	No	Yes	Riparian corridor, stream

watershed-scale management within its approach to designing restoration projects (Table 5). This is indicative of the overall thinking in the field of stream restoration, which focuses on restoration of the streams themselves or the streams and their adjacent riparian corridors rather than on the entire watershed (Simon et al. 2011). This approach is flawed, as streams cannot be separated from their watersheds (Apfelbaum and Haney 2010). Consideration of streams and watersheds as a single unit leads to the realization that watershed restoration encompasses restoration of the terrestrial and aquatic ecosystems within the watershed. The increased complexity of working at the watershed scale has likely led to the focus on riparian corridors and individual streams. It is important to realize that watershed-scale restoration does not require implementing restoration practices within the entire watershed. Instead, watershed-scale restoration involves developing restoration designs after examining conditions and opportunities within the uplands, riparian corridors, and streams within the entire watershed. From this perspective, watershed-scale restoration would encompass the use of targeted implementation of restoration practices to yield the greatest benefits.

The Treatment Train Approach was developed as a way of managing stormwater in urban developments in the Midwest, but this framework transfers well to agricultural watersheds because of its focus on reducing hydrologic transport time and the uptake of sediments, nutrients, and other contaminants. The Treatment Train Approach is the only approach that would apply restoration practices in all three components of a watershed (i.e., uplands, riparian corridors, streams) (Table 5). The Rosgen Natural Channel Design approach falls short of being a holistic approach because it recommends either upland management through altered land-use practices *or* natural channel design that leads to the alteration of the riparian corridor and streams (Rosgen 2007). The Riparian Management System and the Building-Block Model do address the common problem of riparian and instream habitat degradation in many agricultural streams, but the omission of upland practices is problematic because it fails to address the source of ecosystem degradation (agricultural land use). Naturalization, which focuses on setting restoration targets for channelized streams in light of human usage of streams instead of historical conditions, is a broad enough framework that it could conceivably result in a holistic watershed-scale restoration design. However, the Naturalization framework is not applicable to all degraded agricultural watersheds because it was developed to provide alternative management options specifically for channelized agricultural watersheds (Rhoads and Herricks 1996). In summary, we identified the Treatment Train Approach as the most applicable theoretical framework for designing future watershed restoration projects because (1) it involves the use of created, restored, or existing habitats throughout the watershed to assist with addressing the impacts of habitat loss as a result of agriculture; (2) it promotes the use of created, restored, or existing habitats to address the hydrological and chemical impacts of agriculture; (3) the use of different habitat types (forest, prairies, and wetlands) as part of the framework implies the need for multiple types of restoration practices; and (4) it is compatible with the hillslope framework used by the NRCS, where practices located on the top of a hill function to reduce surface runoff flow and capture pollutants at the source and practices located at the bottom trap and treat surface runoff or drainage flow.

USING THE TREATMENT TRAIN APPROACH FOR DESIGNING FUTURE WATERSHED RESTORATION STRATEGIES

Application of the Treatment Train Approach for designing agricultural watershed restoration strategies would first involve identifying and grouping restoration practices by their potential use in different locations within a watershed or hydrologic zone (e.g., upland habitat restoration, reducing surface runoff, reducing subsurface runoff, riparian corridor restoration, or instream and channel restoration). The NRCS currently groups its practices as avoiding, controlling, and trapping practices (Figure 30). Although the Treatment Train terminology is not specifically used by the NRCS, these three categories are similar to the three watershed or hydrologic zone groupings that are the basis of the Treatment Train Approach (Miller et al. 2012). We group potential restoration practices below and discuss in each section how the Treatment Train Approach leads to a more holistic consideration of potential restoration practices to use when designing future watershed restoration strategies. Although a wide range of potential restoration practices are discussed, we are not suggesting that all practices need to be part of every project. That approach would result in unfocused watershed restoration projects characterized by structure mania (Rosgen 1996). We cannot emphasize strongly enough that the selection of restoration practices depends on the goals of the project. The development of specific goals before beginning the restoration project is critical for ensuring success and allocating limited fiscal and physical resources toward high-priority objectives. Involving all watershed stakeholders in the development of

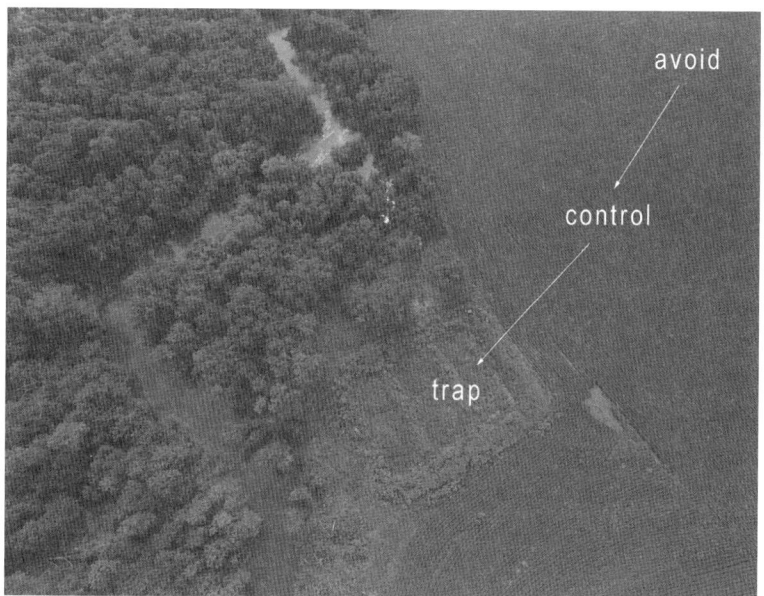

FIGURE 30. The Treatment Train Approach shown on a hillslope in a rural watershed. A treatment wetland is shown capturing runoff from an upstream farm located in Martin County, Minnesota. The NRCS uses the categories of avoiding, controlling, and trapping to characterize agricultural management practices. Practices on the top of the hillslope, such as conservation tillage, avoid loss of soil and nutrients, while practices such as grassed swales control the movement of water and pollutants downstream. Trapping practices capture and treat water before it discharges downstream. In the riparian zone, there is often greater opportunity for ecological restoration practices such as wetland restoration and riparian forest buffers.

the restoration goals is critical for success as well. One key lesson from recent evaluations of the influence of agricultural conservation practices on channelized headwater streams (i.e., agricultural drainage ditches) in the Midwest is that a broader watershed approach is needed, one that addresses agricultural impacts with multiple types of practices (Smiley et al. 2011). Therefore, as highlighted by the Treatment Train Approach, we recommend that future watershed restoration projects in the Midwest be designed to meet their goals using multiple types of restoration practices implemented within all three components of agricultural watersheds (i.e., uplands, riparian corridors, streams) (Figure 28).

Upland Habitat Restoration and Management

Past watershed management approaches in the uplands have focused on reducing sediment, nutrient, and pesticide loss. Typical practices used to achieve this goal include reduced tillage, no-till tillage, crop rotation, cover crops, and contour plowing. Much of the early soil conservation work following the creation of the NRCS in the 1930s focused on erosion control (Argabright et al. 1996). Reducing soil loss is important because it helps maintain long-term soil productivity. From a watershed perspective, reducing sediment, nutrient, and pesticide loss from the uplands is important because it reduces the amount of agricultural contaminants delivered downstream. Federal and state cost-share support for these conservation practices is often available to encourage implementation by private landowners. It is also likely that implementing these practices within the entire upland area in a watershed might not be necessary, as modeling efforts suggest that targeting hot spots within a watershed may significantly reduce agricultural contaminants (Walter et al. 2007). Conservation practices are not traditionally viewed as restoration practices, because they are not intended to increase biodiversity or create specific habitat types. Conservation practices that are effective at reducing sediment, nutrient, and pesticide loading to streams within agricultural watersheds contribute to addressing degraded water chemistry conditions caused by agriculture.

The Treatment Train Approach would lead one to seek out potential opportunities for restoring upland habitats. These opportunities can be identified through a watershed assessment of yield productivity and marginal croplands. Grassland and prairie restoration in the uplands are particularly important for ecosystem services associated with wildlife habitat required by waterfowl, grassland bird, and insect diversity. Although opportunities for large-scale retirement of cropland will be limited, simply assessing the potential areas and exploring the opportunities is important. Cropland retirement has been used in past watershed management approaches. In the 1930s, NRCS efforts in erosion control focused on removing the most erosion-prone and marginal farmland from production and converting it to grassland habitat. Recently, conversion of farmland to grassland habitat has occurred through the Conservation Reserve Program on over 75,000 square kilometers of farmland throughout the United States (FSA 2011). However, the Conservation Reserve Program is not a permanent cropland retirement program, which means these planted grasslands could be put back into agricultural production. Enrollment in this conservation program in the Midwest has decreased by 17 percent from 2007 to 2011 (FSA 2012). Decreases in enrollment are likely due to crop prices and the increased demand for corn (*Zea mayes*) for ethanol production (Secchi and Babcock 2007; Lubowski et al. 2008; Simpson et al. 2009; Classen et al. 2011). USDA commodity programs that provide farmers support for crop insurance, market loans, and disaster assistance also play a role in decreasing Conservation Reserve Program enrollment because these commodity programs provide greater benefits with increased amounts of cropland (Lubowski et al. 2008; Classen et al. 2011).

The few remaining prairie and wetland habitats within the uplands need to be protected because remnant habitats are scarce in midwestern agricultural watersheds. Thus, identifying these habitats and highlighting to landowners and operators their ecological importance and the ecosystem services they provide can be an important part of watershed restoration efforts. Simply ensuring that landowners and operators refrain from eliminating remnant habitats is not typically considered ecological restoration. Given the scarcity of remnant habitats in agricultural watersheds, protecting these habitats should be part of all watershed restoration efforts (Dopplet et al. 1993). One advantage is that simply protecting remnant habitats does not require the farmers to change their cropland management strategies. These remnant habitats and their biota would benefit from increased protection by planting buffers around the remnant habitats and habitat creation efforts that link them with other remnant habitats or increase their size (Dopplet et al. 1993; Kingsbury and Gibson 2002).

The use of perennial crops such as switchgrass (*Panicum virgatum*) or agroforestry crops may also be considered within the Treatment Train Approach because they provide alternative land cover in the uplands. Perennial crops and other types of alternative crops maintain soil productivity, reduce the frequency of soil tillage, reduce reliance on monocultures, reduce pesticide usage, and reduce energy inputs while increasing carbon storage (Jarchow and Liebman 2011).

Reducing Surface Runoff

Surface runoff occurs after upland soils become saturated with rainwater or rainfall exceeds soil infiltration capacity and the excess water flows over croplands to the numerous swales, drainage ditches, and headwater streams within agricultural watersheds. As surface runoff travels over croplands, it picks up sediments, nutrients, and pesticides. After it enters the swales, drainage ditches, and headwater streams, it is transported downstream to larger streams and rivers within the watershed. The goal of water management in the Midwest from an agricultural perspective is to remove excess water from fields quickly to prevent the detrimental impacts of soil saturation on crop growth. Agricultural drainage enables the soil to warm up sooner and prolongs the growing seasons in northern climates (Wright and Sands 2001). Therefore, the challenge for agricultural watershed restoration is to find ways to slow down surface and subsurface water flow, provide water storage, and remove sediment, nutrients, and pesticides through conservation and restoration practices that are compatible with current agricultural practices, and therefore more likely to be adopted by farmers (Nassauer et al. 2007). Here we focus on potential surface runoff practices capable of altering surface water flow and reducing the export of sediment, nutrients, and pesticides. The Treatment Train Approach helps restoration designers focus on identifying such practices to use within the uplands, riparian corridors, and streams.

Agricultural conservation practices that increase infiltration and reduce surface runoff in the uplands should be an integral part of any watershed restoration effort when the goal is to reduce non-point source pollutants and decrease hydrological variability in agricultural streams. Conservation practices such as no-till farming, conservation tillage, and terraces are effective erosion-control practices in part because they promote infiltration, increase soil organic carbon, and reduce surface runoff. Wetland restoration or creation within uplands or riparian corridors to capture surface runoff can also reduce surface runoff, remove agricultural contaminants from surface runoff, and provide habitat for wetland-dependent plants and animals (Galatowitsch and van der Valk 1994). Even low-quality existing wetlands may provide the same ecosystem services (Galatowitsch and van der Valk 1994). Depending on the position of restored, created, and existing wetlands within the watershed, they may serve different functions. Floodplain or riparian wetlands are useful for capturing edge-of-field sediment, nutrients, and pesticides, delaying flood flows, and reducing peak flood flows downstream (Mitsch 1992). Upland wetlands are more likely to serve as groundwater recharge zones, as water tends to infiltrate higher up the hillslope and discharge at lower landscape positions into streams and wetlands (Winter 1989). Much of the past wetland restoration work in the Midwest has focused on providing waterfowl habitat. Future wetland restoration and creation efforts undertaken with multiple objectives to reduce surface runoff, capture agricultural contaminants for water quality improvement, and provide habitat for aquatic life are most likely to assist with watershed restoration efforts.

Reduction of surface water flow and removal of nitrate and other dissolved pollutants is hindered by channelization for agricultural drainage. Channelization-induced increases in surface flow result from the construction of enlarged channels designed to contain large flood events and the straightening and simplification of the channel through the removal of instream wood and other retentive structures. Channelization-induced increases in surface flow and the associated geomorphological changes in turn increase the export of nutrients and other agricultural contaminants. Instream drainage water management structures or small impoundments designed to reduce surface flow can reduce nutrient, pesticide, and sediment export from channelized headwater streams (Strock et al. 2007; Kroger et al. 2012). However, these practices should only be used within channelized headwater streams that are hot spots for nutrient, pesticide, and sediment transport because they will alter the habitat and potentially lead to habitat fragmentation. Increased sinuosity leads to greater water retention times and greater uptake of nutrients within streams. Thus, restoration practices that reestablish meanders within channelized headwater streams may assist with reducing surface discharge and downstream transport of agricultural contaminants. Reducing the frequency of channel maintenance in channelized streams may also lead to reduced nutrient and pesticide transport by promoting the establishment of aquatic vegetation and other retentive features (Bouldin et al. 2005; Lester and Boulton 2008).

Unchannelized headwater streams are rare within agricultural watersheds in the Midwest (Mattingly et al. 1993). Headwater streams tend to be more efficient at removing nitrogen than larger rivers (Alexander et al. 2000; Peterson et al. 2001), and less disturbed agricultural streams have a greater ability to reduce downstream nitrogen transport (Schaller et al. 2004). Thus, restoration efforts that protect unchannelized streams should assist with reducing downstream export of agricultural contaminants.

Reducing Subsurface Runoff

Subsurface flow in agricultural watersheds consists of natural groundwater flow and subsurface drainage flow. We focus on subsurface drainage flow here due to the large-scale hydrological impacts of subsurface drainage pipes in the Midwest. Simply put, subsurface drainage pipes reduce water storage capacity and increase subsurface flow from the uplands to the riparian corridors and streams. Subsurface drainage has decreased the amount of upland field and gully erosion, but it has increased the volume of water entering Midwest streams (Lenhart et al. 2011) and has resulted in more channel erosion. The Treatment Train Approach would lead restoration designers to seek opportunities for increasing water storage and reducing subsurface flow to address these hydrological modifications.

Implementing practices that alter subsurface drainage within Midwest agricultural watersheds will be a challenge because farmers rely on subsurface drainage pipes to ensure production. Simply modifying the subsurface drainage network by increasing the spacing and reducing the depth of the drainage pipes would reduce the volume of water and nitrogen inputs into streams (Wright and Sands 2001). New practices such as drainage water management, routing subsurface runoff into treatment wetlands, and dissipation throughout riparian corridors have been developed to mitigate the impacts of increased subsurface drainage runoff and to reduce the loadings of agricultural contaminants. Drainage water management uses structures that hold back subsurface flow in subsurface drainage pipes in the winter and early spring before the crops are planted to create anoxic conditions in upland agricultural fields. This practice reduces the amount of runoff, nitrate loading, and dissolved phosphorus. Initial assessments of drainage water management in northwest Ohio indicate this practice slightly increased corn and soybean (*Glycine max*) yields (Ghane et al. 2012). Drainage water management is compatible with existing agricultural practices, but it's not likely to be widely adopted without economic incentives and outreach efforts to overcome concerns about the potential effects on crop yields. Additionally, drainage water management typically involves unrestricted drainage in the spring following planting, fertilization, and pesticide application, and is not likely to reduce the typical late spring–early summer increases in nutrient and pesticide concentrations. In the summer and the autumn, partial restriction of drainage with the control structures is promoted as a way of providing water for crops during periods of drought and limited rainfall.

Subsurface flow typically bypasses riparian corridors in agricultural watersheds in the Midwest (Smiley et al. 2011). Routing subsurface drainage flow into treatment wetlands created to filter out nutrients and dissipate pesticides prior to their entry into surface waters has been promoted to address chemical degradation (Osborne and Kovacic 1993; Smiley et al. 2011). Creation of wetlands for treatment of subsurface drainage runoff also provides flood storage and habitat for wetland-dependent animals. In regions where sub-irrigation is used, agricultural water recycling systems have been designed that create wetland habitat, treat agricultural runoff, provide sub-irrigation water for crops, and increase crop yields (Smiley and Allred 2011). Practices that dissipate subsurface runoff through riparian corridors and prevent the runoff from bypassing these corridors also hold much promise, but they have not been widely used or evaluated. One newly developed practice is the saturated buffer (Jaynes and Isenhart 2014). Saturated buffers involve the use of a water control structure installed in the subsurface tile drain that diverts subsurface runoff laterally into the riparian corridor rather than allowing it to enter directly into the stream. Initial evaluations suggest saturated buffers are capable of reducing nitrate input into agricultural streams (Jaynes and Isenhart 2014). Modifying subsurface drains so that they open into the riparian corridor rather than into the streams has also been recommended as a way of dissipating runoff through the riparian corridor (Vought and Lacoursiere 2010).

Riparian Corridor Restoration

In the Midwest, many riparian corridors have been narrowed and converted from forested to herbaceous riparian habitats, especially along smaller streams. These modifications have resulted in watersheds where most remnant habitats are confined to the riparian corridors because the uplands are dominated by cropland. Watershed management efforts have widely promoted management of riparian vegetation to the agricultural community in an attempt to reduce sediment, nutrients, and pesticide loading to downstream resources. However, riparian corridors provide valuable ecosystem services beyond their ability to reduce loadings of agricultural contaminants, and numerous restoration opportunities exist within these streamside habitats (Smiley and Rumora 2015). The importance of ripar-ian corridors is evident, as four of five possible frameworks applicable for designing watershed restoration strategies recommend the use of ripari-an corridor practices (Table 5). The Treatment Train Approach, with its emphasis on habitat creation and restoration throughout the watershed, will lead restoration designers to design future strategies that incorporate restoration into riparian corridors because future opportunities will most likely occur within this watershed zone.

Many streams possessed forested riparian corridors prior to the estab-lishment of agriculture. Existing forested riparian corridors need to be pro-tected, especially those associated with unchannelized headwater streams. Educational efforts on the ecological value and ecosystem services provided by forested riparian corridors would assist with protection efforts because many in the agricultural community view trees as a threat to agricultural drainage. Educational efforts would also assist with preventing misguid-ed stewardship efforts. For example, we are aware of an instance where a midwestern farmer identified himself as a steward of the land to a visiting group and then described how he had removed all of the trees within a mature hardwood forest riparian corridor and planted a grass filter strip in their place. Protection efforts should also extend to practices that prevent and control invasive plant and insect species that can cause tree mortality (see Chapter 8). These protection efforts contrast with traditional riparian restoration that focuses on planting riparian vegetation, but they are likely to be important components of agricultural watershed restoration strategies.

Practices that increase the width of riparian corridors and shift the ri-parian vegetation from herbaceous to forested cover will contribute to watershed restoration efforts. Currently there are no laws or regulatory standards for maintaining minimum riparian widths on agricultural land in many midwestern states. Increasing riparian width will be difficult, as it will most likely entail relying on farmers to voluntarily take cropland out of production. Farmers are also hesitant to plant woody vegetation or allow woody vegetation reestablishment adjacent to streams because of their focus on agricultural drainage and desire to keep obstacles out of the way of farm machinery.

Planting or passively allowing the reestablishment of woody vegetation in riparian corridors has not been widely explored in the Midwest in the context of watershed management. Instead, grass filter strips have been promoted as an effective method for reducing sediment, nutrient, and pesticide loads, and they have been implemented widely throughout the Midwest (Smiley et al. 2011). Unfortunately, the vast amount of research on this practice has been limited to small field-scale assessments, and there is only a limited amount of information on the impacts of this practice at the watershed scale (Smiley et al. 2011). Limited assessments of the impacts of grass filter strips adjacent to channelized headwater streams in central Ohio suggest that filter strips simply widen the riparian corridors and are not contributing to the restoration of riparian ecosystem structure or function (Smiley et al. 2011; Smiley and Rumora 2015). These findings suggest that grass filter strips need to be combined with practices capable of reducing loadings of agricultural contaminants or improving instream habitat conditions (Smiley et al. 2011; Smiley and Rumora 2015). One such multipractice strategy promoted by others (Isenhart et al. 1997) is the installation of grass filter strips adjacent to forested riparian corridors. This combination has the potential to increase sediment trapping efficiency (Nieber et al. 2011).

Riparian corridors in agricultural watersheds typically lack the wetlands that characterized these ecosystems prior to implementation of agriculture. Thus, wetland creation within riparian corridors provides yet another opportunity to create wetland habitat and increase the physical and biological diversity of riparian corridors. As mentioned previously, these created wetlands can be designed to capture subsurface drainage to improve filtration and provide habitat for aquatic animals. Granted, larger wetlands are more biologically valuable because of their ability to support a greater number of species. But even small ephemeral riparian wetlands that are important for amphibians and reptiles (Kingsbury and Gibson 2002) can contribute to restoration efforts, especially if they can be widely implemented throughout the watersheds.

Instream and Channel Restoration

The Treatment Train Approach will lead restoration designers to consider the use of instream and channel practices to address physical habitat degradation, reduce instream surface flow, and reduce the downstream export of agricultural contaminants. Small-scale restoration projects that implement instream and channel practices are common throughout the Midwest, especially within watersheds containing coldwater streams and trout (Moerke and Lamberti 2004; Alexander and Allan 2006). Use of instream habitat practices for stream restoration was pioneered in the Midwest and later widely applied for salmonid restoration efforts in the Pacific Northwest (White 1996). The lack of a game fishery for warmwater streams has resulted in less emphasis on instream habitat modifications and greater emphasis on improving degraded water chemistry conditions (Alexander and Allan 2006).

Undoing the impacts of widespread channelization and straightening of stream channels for agricultural drainage and bridge maintenance should be a major component of any agricultural watershed project because of the well-documented impacts of channelization on physical habitat and biota (Brookes 1989; Smiley and Gillespie 2010). Many Total Maximum Daily Load plans for midwestern agricultural watersheds require decreases in loadings of agricultural contaminants, the improvement of physical habitat conditions, and subsequently the ecological integrity of channelized streams. This is a significant challenge for the agricultural community, which has limited experience with stream management and places little value on the aquatic life within these streams.

Large wood in streams provides cover for aquatic animals, serves as a habitat-forming feature, promotes nutrient uptake, and facilitates sediment retention (Gregory et al. 2003). Channelization and channel maintenance practices simplify instream habitat when large wood—greater than one meter long and ten centimeters in diameter—is removed. Placement of large wood in channelized streams or policies that reduce its removal have the potential to enhance instream habitat (Lyons and Courtney 1990; Cordova et al. 2007). Particularly, the addition of large wood structures designed to create pool habitat are more likely to have greater effects than large wood structures intended only to provide cover (Smiley and Gillespie 2010).

Streams naturally undergo changes in channel geomorphology through time due to erosional processes that result in channel migration. Excess sediment from field erosion initiated by poor watershed practices has been described as one of the single largest agricultural impacts on streams (Waters 1995). Excess sediment buries coarse-grained spawning beds, changes biogeochemical reactions, leads to eutrophication, and reduces instream biotic diversity. Elevated rates of stream erosion from increased runoff or channel modifications can also export substantial amounts of sediment and phosphorus downstream. Practices that reduce excess sediment loads have the potential to contribute to watershed restoration projects. Numerous bank stabilization projects have been conducted in the Midwest to reduce sediment loss (Moerke and Lamberti 2004; Alexander and Allan 2006). Many projects have implemented engineering practices that impose channel stability and reduce lateral channel migration. Prohibiting channel migration reduces streams' natural heterogeneity, which is critical for maintaining biological diversity. Thus the challenge is to use practices that reduce excess sedimentation and allow lateral channel migration.

Geomorphologists have developed a wide variety of approaches for natural channel design for stream restoration (Skidmore et al. 2001). The term *natural channel design* refers broadly to methods that use geomorphological data from intact, healthy streams or historical stream channel conditions from the study area to guide restoration of channel form and function in degraded streams that have modified geomorphology. Recently, natural channel design has become synonymous with a specific approach pioneered by David Rosgen (Rosgen 1996, 2007). Rosgen's approach was intended to provide an alternative to traditional river engineering, which focused on single objectives related to flood control, drainage, and navigation (Rosgen 1996, 2007). His approach consists of a national channel classification system (Rosgen 1994) and a forty-step procedure for designing stream restoration projects that encompasses the entire design process from developing restoration objectives to post-project monitoring (Rosgen 1996, 2007). Rosgen's approach has been widely adopted and accepted in the United States by consultants and some government agencies despite being heavily criticized by academic and federal geomorphologists (Lave 2009). As noted previously in our evaluation of the five theoretical frameworks, the Rosgen approach does not lead restoration planners to develop a holistic watershed restoration project involving the use of upland, riparian, and stream restoration practices. The Rosgen approach, or modified versions of it, has been used to design stream restoration projects in Wisconsin and Minnesota (Rosgen 2007; see Chapter 5). Other natural channel design approaches have also been used in the Midwest, such as the project conducted on Nippersink Creek, Illinois (Simpson 2008), and The Nature Conservancy's restoration of the Big Darby Creek headwaters in Ohio. Notably, the Nippersink Creek project incorporated a design that combined the use of upland wetland, grassland, and forest restoration, natural channel design, and instream habitat structures (Simpson 2008).

The Treatment Train Approach also leads restoration designers to consider alternative drainage designs for channelized headwater streams, such as the two-stage channel design (see Chapter 2) because of its potential to reduce downstream nutrient transport and benefit biota by providing hydraulic refugia during flooding events within these small streams. Additionally, a more economical option to consider for channelized headwater streams with stable stream banks would be to allow the stream to develop the characteristic two-stage benches on its own (Landwehr and Rhoads 2003; Rhoads and Massey 2012) instead of intensively restructuring the channel.

The widespread use of instream habitat structures and channel practices in small stream reaches has prompted many to discourage their use and instead recommend addressing the larger scale problems that caused instream habitat degradation (Dopplet et al. 1993; Williams et al. 1997; Roni et al. 2008; Bernhardt and Palmer 2011). Yet the high level of physical habitat degradation in the Midwest will often require the use of instream and channel practices combined with riparian corridor and upland watershed practices. We do not recommend the widespread use of these practices for all watershed restoration projects. Instead, judicious use of instream and channel practices can be short-term solutions to initiate the watershed recovery process as the practices implemented in the uplands and riparian corridors mature and exert the needed long-term effects.

Benefits and Limitations of the Treatment Train Approach

The primary benefit of the Treatment Train Approach is that it forces those designing future restoration strategies to focus on the entire watershed and the appropriate siting of different practices within different watershed zones to achieve the desired restoration goals. Consideration of the entire watershed is important because previous watershed restoration strategies have focused on the riparian corridors and individual stream sites but have failed to address the root cause of degradation: agricultural land use.

In many midwestern agricultural watersheds, the conversion of cropland to forest or prairie habitat is not likely to occur on a large scale. The Treatment Train Approach is invaluable in addressing this restoration constraint by prompting designers to consider alternative ways to increase hydrologic storage and residence time within the watershed and decrease loading of agricultural contaminants. The Treatment Train Approach also enables selection of restoration practices that are most appropriate for the watershed location, and it will assist with prioritization of restoration efforts within watersheds. The entire watershed focus makes the Treatment Train Approach more suitable than the other frameworks (Table 5), which focus on restoration of riparian corridors and streams instead of the entire watershed. Additionally, the Treatment Train Approach focuses on improving ecosystem processes that should result in improved ecosystem services of flood reduction, improved water quality, and habitat for terrestrial and aquatic biota through the creation and restoration of habitats within the watershed. This process-oriented approach highlights how ecological restoration can be used to meet restoration goals established using ecosystem services and watershed management approaches.

The Treatment Train Approach does have some limitations. The focus of the framework on hydrologic transport and water quality improvement might imply to some that physical habitat modifications of the riparian corridors, channels, and instream habitat are not necessary. Restoration of agricultural watersheds requires more than simply addressing water quality issues, and improvement of physical habitat will be a necessary component of many future watershed strategies (Isenhart et al. 1997). The Treatment Train Approach may also be interpreted to imply that any created or restored habitat is simply a sacrifice zone for reducing hydrologic transport and for water quality treatment. For example, small first-order headwater streams and riparian wetlands may be valued only as a strategy to reduce the loading of agricultural contaminants into downstream areas. These small, embedded aquatic ecosystems are valuable ecosystems unto their own right, and their use as sacrifice zones may reduce their biodiversity and aesthetic value. The key is to understand the implications of this trade-off and attempt to avoid restoration plans that consider all small habitats as sacrificial zones. The Treatment Train Approach is a valuable tool for selecting restoration practices and deciding where to place them within the watershed. Watershed restoration strategies encompass more than just selecting and identifying where to place restoration practices; they also involve promoting stakeholder involvement and pre- and post-evaluation of the watershed restoration strategy. Although the Treatment Train Approach does not explicitly incorporate these components of watershed restoration planning, it can be used in conjunction with them.

CONCLUSIONS

The approaches and strategies we discuss here are based on the best available science and professional experience. Unfortunately, we are unable to identify which watershed restoration strategies or practices will be the most effective because only a limited number of agricultural watershed restoration projects have actually been conducted within the Midwest and evaluation of these projects is lacking. Recent surveys of stream and river restoration practitioners suggest that very little written documentation evaluating stream and river restoration projects is available (Bernhardt et al. 2007). Unpublished evaluations contribute little to furthering the practice and science of watershed restoration. It has been noted that some of the best thinking in ecological restoration is not well represented in the scientific literature because practitioners do not regularly publish in scientific journals (Jordan 2003). We need more quantitative evidence on which practices and watershed restoration strategies are the most effective. Ultimately, the biota and the people living in these degraded watersheds are the ones that suffer because of the lack of guiding knowledge to develop and implement effective agricultural watershed restoration projects.

While it is common knowledge that most river and stream restoration projects are small-site projects, there is encouraging evidence that an increasing number of these projects are being implemented in context of larger scale watershed restoration plans (Wheaton et al. 2006; Alexander and Allan 2007). The challenge lies in determining how to further promote the implementation of watershed restoration projects. In the Midwest, historical stream management projects were initiated by individual landowners, and future restoration efforts will also be dependent on the willingness and initiative of individual landowners. Watershed restoration projects developed under the economic and social constraints imposed by agriculture are likely to be more widely accepted by the agricultural community and hence more frequently implemented. Economics play a key role in motivating midwestern farmers to participate in watershed restoration projects (Johnson et al. 2012). Providing funding support for implementing restoration and alternative agricultural practices through programs such as the NRCS Environmental Quality Incentives Program will likely increase farmer participation in future watershed restoration projects. Greater outreach efforts that promote an understanding that ecological restoration can complement agricultural production should reduce the cultural barrier to watershed restoration in the Midwest. In Minnesota, the development of Watershed Restoration and Protection Strategies (WRAPS) promises to better integrate restoration projects into watershed plans (Anderson et al. 2016). Many plans incorporating these strategies have been developed and serve as guidance documents (MPCA 2014), but these plans have not been scientifically evaluated as of 2017.

The future will surely bring more challenges for restoring agricultural watersheds. Climate change will most likely continue to alter temperature and precipitation regimes in the region (see Chapter 7), which will in turn lead to changes in agricultural practices and hydrological regimes in agricultural watersheds. These changes will likely intensify the impacts of agriculture on the environmental health and biota of midwestern agricultural watersheds. Future climate change may also exacerbate the existing invasive species problems within the region (see Chapter 8) and create new threats. The impending threats from climate change make it more critical than ever for us to implement effective watershed restoration projects in the Midwest.

Restoration of agricultural watersheds in the Midwest may seem impossible to some, especially in the face of new and emerging threats. Skeptics and pessimists will argue that watershed restoration is too difficult, too time consuming, and too expensive to attempt. In contrast, we feel that despite the challenges, it is possible and necessary to move from site-level to watershed-scale restoration efforts to repair degraded agricultural watersheds and enhance the valuable ecosystem services they provide. Our optimistic viewpoint is supported by examples of midwestern farmers (Jackson and Jackson 2002) who have successfully adopted alternative, sustainable farming practices that if implemented on a large spatial scale have the potential to contribute greatly to future watershed restoration efforts. Our perspective is also supported by the land management expertise available within universities, state agencies, federal agencies, and the numerous private ecological restoration firms in support of agricultural watershed restoration. Our optimism is further supported by past collaborations among these expert groups in the region, such as those that occurred as part of the development of many of the WRAPS for agricultural watersheds of Minnesota. For us, pondering the union of restoration and agriculture leads to the realization that there are better options and that it is no longer necessary to accept the degradation of the Midwest's agricultural watersheds as a foregone conclusion (Jackson and Jackson 2002).

CHAPTER TEN

BUILDING ON DUAL LEGACIES OF ECOLOGICAL RESTORATION AND ECOSYSTEM DEGRADATION

PETER C. SMILEY JR. AND CHRISTIAN LENHART

Ecological restoration in the Midwest is a diverse professional discipline with a long history and a promising future. To share its unique attributes, we compiled essays from regional experts that discuss the history, restoration case studies of aquatic and terrestrial ecosystems, and future restoration issues and challenges pertinent to ecological restoration in the region. In doing so, we also illustrated how the legacy of the science and practice of ecological restoration forms the foundation for current and future restoration efforts in the region. Our objective for this concluding chapter is to provide a summary of the primary lessons learned from the three sections of the book and then describe our vision for the future of ecological restoration in the Midwest.

Our review of the history of ecological restoration in the Midwest (see Chapters 1 and 2) highlights the shared historical link between ecology and ecological restoration in the region. Prior to its conceptualization, the foundations of ecological restoration were laid by the historical precursory efforts conducted in the late 1800s by biologists, ecologists, landscape architects, and natural resource managers. In some cases, these historical precursory efforts resulted in long-lasting accomplishments (e.g., the protection of important natural areas; modern restoration projects; the development of methods to evaluate restoration success; and the formation of state, national, and international organizations) that continue to contribute to ecological restoration within the region. Additionally, a number of important theories (e.g., succession, relationship between diversity and stability, trophic cascades, alternative stable states) and practices (e.g., reference ecosystems, oak savanna restoration techniques, prescribed fire, biomanipulation, two-stage channel design, dam removal) were developed in the Midwest or their use was further refined in the region. These theories and restoration practices are applicable to the field of ecological restoration internationally and highlight the contributions of individuals and organizations from the Midwest to the global restoration community. Previous narratives of the history of ecological restoration (Jordan and Lubick 2011) focus on terrestrial ecosystems and imply that the Midwest only contributed to furthering our understanding of terrestrial ecosystem restoration. In contrast, our review of the history of the field illustrates that individuals and organizations from the region were pioneers in both aquatic and terrestrial ecosystem restoration. This is an important insight, because it expands our understanding of the history of ecological restoration in the region.

A common theme in our review of the history of ecological restoration is the importance of integrating science with practice. The provided examples are those with direct and indirect implications. Clear examples of the direct implications are the development of oak savanna restoration techniques through the process of learning by doing (see Chapter 1) and the concurrent development of the trophic cascade theory, alternative stable state theory, and the practice of biomanipulation within lakes (see Chapter 2). Examples of indirect implications include the ecological evaluation of the Chicago Drainage Canal's impacts on the Illinois River and the documentation of the relationships between biological diversity and physical habitat diversity in midwestern streams. Both led to the development of water quality indexes than can be used to evaluate the success of future restoration efforts (see Chapters 1 and 2).

The presented case studies on prairie restoration (see Chapter 3), floodplain wetland restoration (see Chapter 4), stream restoration (see Chapter 5), and urban ecosystem restoration (see Chapter 6) provide site-specific and ecosystem-specific details on restoration efforts in the Midwest. These case studies build on our regional legacy in restoration and highlight how ecological restoration goes beyond simply establishing historical community structure to encompass the reintroduction of essential ecosystem processes (e.g., fire, flooding, natural flow regimes) and habitat manipulations (e.g., geomorphology, soil characteristics, invasive species) to repair ecosystem structure and function within damaged aquatic and terrestrial ecosystems. Also common among these case studies of successful restoration efforts is the collection of information before, during, and after for developing the restoration objectives and design, evaluating restoration outcomes, and determining management actions needed post-restoration.

Cumulatively, the presented case studies represent an evolving knowledge and appreciation among restoration practitioners, natural resource managers, and the public regarding the importance of terrestrial and aquatic ecosystem functions and services. In the 1960s, only a few people realized the importance of burning prairies, flooding productive agricultural lands for floodplain wetland restoration, reestablishing sinuosity in channelized streams, and restoring forests within urban areas. Back then, these were viewed as fringe ideas held by radical environmentalists. Recently, these ideas have gained broad acceptance among individuals and organizations involved with conservation and restoration in the Midwest and elsewhere.

The future issues chapters focus on restoration challenges involving climate change (see Chapter 7), invasive species (see Chapter 8), and agricultural land use (see Chapter 9). These restoration issues are complex because they involve a high degree of uncertainty and encompass multiple ecosystem types in the Midwest. These three chapters describe the use of novel and untested approaches for designing future restoration efforts to address these restoration challenges. These chapters also build on the rich legacy of restoration in the Midwest and represent the next step toward achieving ecological sustainability. It is our hope that this information will inspire others to use and evaluate these approaches, then share their findings. The novel approaches discussed were developed in part based on predictive models or theory and are reliant on adaptive restoration to enable adjustments to changing environmental conditions and unexpected restoration outcomes. As a result, these chapters also highlight the importance of linking theory and practice in the development of novel restoration approaches.

VISION FOR THE FUTURE OF ECOLOGICAL RESTORATION IN THE MIDWEST

The Midwest has a legacy of innovative research and implementation of restoration practices in ecological restoration (see Chapters 1, 2, 3, and 9).

The Midwest also has a legacy of ecosystem degradation resulting from agriculture, urbanization, industrialization, and invasive species (Ashworth 1986; see Chapters 6, 8, and 9). Therefore, the challenge for future restoration efforts in the Midwest will be to address historical impacts within the context of future threats from climate change, the continued onslaught of invasive species, continued increases in agricultural and urban land uses, and yet unidentified stressors. In this section, we describe our vision for the future of ecological restoration in the Midwest based on the lessons learned and links between the previous chapters.

Our review of the history of ecological restoration in the Midwest indicates that advances were made most effectively when theory (i.e., a set of guiding principles with a mixture of physical science, ecology, and management precepts) was integrated with practice in restoration efforts. Prominent examples include the Curtis Prairie research studies documenting the importance of fire for prairies (see Chapter 3) and the lake research studies that fueled the development of the trophic cascades concept, alternative stable state theories, and the practice of biomanipulation (see Chapter 2). If we take a broader perspective and consider the integration of science (i.e., the use of the scientific method or information obtained with the scientific method) and practice, then all contributing chapters provide examples of the integration of science and practice in ecological restoration in the Midwest. These examples are wide ranging and include the collection of information to document the delisting of the Ashtabula River Area of Concern during the restoration process (see Chapter 6); the use of the reference ecosystem approach and the pre-restoration collection of information to guide the design of channelized streams and impounded streams in Minnesota (see Chapter 5); the pre- and post-restoration collection of information to guide restoration on the former Acadia golf course in northeastern Ohio (see Chapter 6); and the extensive development of restoration criteria to be evaluated before, during, and after restoration as part of the Emiquon restoration project in Illinois (see Chapter 4). These examples lead us to the realization that the integration of science and practice may be more important to the field than the integration of theory and practice. We are not implying that the integration of theory and practice is not important for the continued maturation of the field of ecological restoration. We instead envision that placing a greater emphasis on integrating science and practice will lead to more collaboration between scientists and practitioners, which in turn should increase the number of research projects that provide practical information needed by restoration practitioners to produce restoration outcomes that address ecological and societal needs.

The importance of adaptive restoration (i.e., adaptive management as applied to restoration practice) for the success of future restoration efforts is highlighted within the case study chapters and the future issues chapters. While information collected as part of adaptive restoration may not constitute a research project, it does represent the collection of information that enables interpretation of the restoration outcome. In this sense, it constitutes a compressed scientific evaluation conducted on a shorter time frame, reduced spatial scale, and with less funding than a full-scale scientific evaluation. If these compressed and less formal evaluations are designed using a priori questions of interest that function similar to hypotheses, then they become more similar to a formal scientific evaluation. By viewing compressed evaluations as a form of applying science to practice, it becomes clear that increasing the use of adaptive restoration will contribute to increasing the integration of science and practice in ecological restoration in the Midwest.

Adaptive restoration is commonly associated with the collection of information after the implementation of a restoration project. In contrast, all of the contributing chapters provided examples that collectively highlight the importance of collecting information before, during, and after restoration to evaluate restoration outcomes. In essence, this is the adaptive restoration model we recommended, using selected objectives and questions developed a priori.

However, the examples provided in the book likely represent best-case scenarios, as many projects do not include scientific and objective evaluations of restoration outcomes. The integration of science and practice is a long-standing problem that the natural resource management fields and other applied fields such as engineering have struggled with (Cabin 2011). Thus, ecological restoration is not alone in its struggle to integrate science and practice. We emphasize that the responsibility for this integration falls to both scientists and practitioners. It is a challenge that will require scientists and practitioners to work together to find a solution.

We envision that restoration in the Midwest in the future will need to capitalize fully on opportunities to upscale restoration efforts from the individual site level to the regional level. The importance of upscaling restoration efforts is highlighted frequently within the contributing chapters. The cumulative impacts of ecosystem degradation and future threats from climate change, changing land use, invasive species, and unknown threats require that we move beyond the era of restoration simply as a response to site-level ecosystem degradation to a new era of proactive restoration design and placement. Upscaling to large spatial scales will be a challenge because the majority of the land use in the Midwest consists of privately owned agricultural and urban land where the degree of ecosystem degradation is the greatest but the restoration opportunities are the most limited.

The contributing chapters also indicate that restoration efforts in the region consist of a diversity of efforts. They range from ideal ecocentric restoration efforts conducted to restore ecosystem structure and function to limited restoration efforts conducted primarily to achieve anthropogenic objectives. Likely it will not be logistically feasible to upscale the ideal restoration efforts, because these projects are intensive in terms of labor and costs. Therefore, one potential strategy is to link restoration efforts that only provide limited restoration benefits to more intensive efforts within a large spatial area to achieve greater cumulative benefits. The vision for future floodplain wetland restoration along the Illinois River (see Chapter 4) is an excellent example of this, as it describes the implementation of a range of floodplain restoration efforts involving limited, partial, and open hydrologic connections over a large spatial area along this major midwestern river.

Another potential upscaling strategy is to implement less intensive restoration efforts with primarily anthropogenic objectives over large spatial scales. The Iowa Integrated Roadside Vegetation Management Program (see Chapter 3) is an excellent example of this, as it describes how planting native prairie plants along roadsides in Iowa was used to reduce the costs associated with weed control. The limited restoration benefits are counteracted by the ability of this program to be applied at large spatial scales. There is the concern that restoration efforts with solely anthropogenic objectives may be cynical applications of the moniker *restoration*. Unfortunately, past management efforts in the region fall into this category, particularly regarding stream ecosystems (see Chapter 5). It is important that designers of restoration efforts with solely anthropogenic objectives clarify their sincerity by explicitly describing how their projects will repair or reestablish essential ecosystem functions. Doing so should separate sincere from cynical attempts.

We also feel it is important to capitalize on opportunities to facilitate the collaboration between multiple government, university, and nonprofit organizations and private companies as part of future restoration efforts in the region. The importance of such collaboration was highlighted frequently in the contributing chapters, and cooperative efforts provide another way of upscaling future restoration efforts within the region. The U.S. Environmental Protection Agency's Great Lakes Restoration Initiative is a good example of a recent coordinated effort among multiple federal agencies to support state and individual restoration efforts. Other good examples of coordinated partnerships include the partnership formed to create one of the largest floodplain restoration efforts in the Midwest, the Emiquon Complex (see Chapter 4); the restoration strategy developed and implemented to increase the resilience of northern forests to future climate change (see Chapter 7); and the cooperative weed management areas formed to address the invasive species problem in the region (see Chapter 8). Particularly important for facilitating partnerships is for the restoration community to identify what goals they have in common with other natural resource agencies and organizations in the region. Realistically, it will not be possible for every restoration effort to involve a high degree of coordination among multiple organizations. However, it is certainly feasible for individuals and organizations to ensure that regional partners are aware of where new restoration efforts are being made so they can link them when possible to achieve greater benefits. Thus, coordination begins with the planning process for implementing individual smaller-scale restoration efforts by considering the region outside of the proposed restoration site.

Future restoration efforts also need to capitalize on the opportunity to involve volunteers. Volunteers have been part of historical and current restoration efforts in the Midwest (Chapters 1, 2, 6, and 8). They represent a source of low-cost labor that project managers can use to maximize their funds and efforts. More importantly, volunteers represent a valuable educational and public relations opportunity, as they can increase the public's understanding of an environmental problem and awareness of the restoration effort within their local community. Volunteering can also provide an opportunity for individuals to cultivate a relationship with nature. Ultimately, the use of volunteers is an avenue for increasing public support of large-scale ecological restoration efforts throughout the region.

Unification of the field of ecological restoration will also be a critical component to ensure its future success in the Midwest. Ecological restoration needs to move away from the divisive dichotomies of science and practice, forest restoration and lake restoration, plant restoration and animal restoration, and numerous other divisions existing in the field to unify the discipline with a common goal: restoration of degraded, damaged, and destroyed ecosystems. Unfortunately, differences among subspecialties are represented by the different definitions of restoration used by those working in stream and lake ecosystems versus those working in terrestrial and wetland ecosystems. The divisive dichotomies that currently exist within ecological restoration may in part be due to the discipline's youth, as ecological restoration as a formal professional discipline is only twenty-nine years old (see Chapter 1).

The field has matured since its beginnings in 1988, and exciting evidence of its maturation is the Society for Ecological Restoration's recent development of international professional standards for the practice of ecological restoration (McDonald et al. 2016). We hope a shift toward unification with a renewed focus on a common goal will be part of the future maturation of the field. The irony is not lost upon us that the conceptualization of restoration ecology (Jordan et al. 1987) was in part an attempt to unite scientists and practitioners by drawing attention to the value of restoration practice as a technique for research. In the last twenty-nine years, much has been written regarding the value of restoration practice as a technique for research. We feel the next phase of unification needs to emphasize that the common goal of the interdisciplinary field of ecological restoration is the repair of damaged, degraded, and destroyed ecosystems.

Aldo Leopold noted many years ago that complex problems require the integration of knowledge from across a broad array of disciplines (Cabin 2011), and designing effective restoration efforts will also require inputs from multiple disciplines (Miller et al. 2017). The many types of restorationists (e.g., scientists, practitioners, engineers, ecologists, soil scientists, climatologists, etc.) in different subspecialties of ecological restoration (stream restoration, lake restoration, prairie restoration, forest restoration, invasive species, etc.) should not be strangers to each other. We have a common goal that results from a shared love of natural ecosystems—a desire to repair damaged, degraded, and destroyed ecosystems—and we need to work together to achieve this common goal. Interdisciplinary professional meetings such as the Annual Meetings of the Society for Ecological Restoration's Midwest–Great Lakes Chapter represent great opportunities to bring together a diverse group of individuals and organizations interested in ecological restoration in the Midwest to enable them to discover a common goal. Differences in opinion will always be present in every professional discipline, but these differences need to be viewed as aspects of the field that need further clarification and examination so the field can progress. Particularly, a unified front in the field of ecological restoration is critical for enabling policy makers, industry leaders, sociologists, and the general public to understand the need for ecological restoration and the great potential it has for improving life for humans, for flora and fauna, and for the ecosystems that all life in the Midwest depends upon.

In conclusion, the final critical component for future ecological restoration in the Midwest is optimism. Ecological restoration is an optimistic endeavor because it can address the impacts of human-induced ecosystem damage and destruction (Jordan 2003, Palmer et al. 2006, Cabin 2011, Perring et al. 2015). It is based on the premise that it's possible to repair damaged, degraded, and destroyed ecosystems. Ecological restoration promotes optimism for scientists and ecologists because it gives them an opportunity to apply their knowledge and research findings toward this goal. It promotes optimism for practitioners because it provides an opportunity to creatively design and implement projects to repair damaged ecosystems. Ecological restoration promotes optimism for the public because it enables them to engage with nature in a deeper way than they would in their everyday lives, and it provides a way for concerned individuals and organizations to contribute actively to addressing the impacts of environmental degradation and destruction (Jordan 2003). To pessimists, many of the restoration challenges in the Midwest seem insurmountable. We, the restoration community, need to be optimistic and envision that along with our partners from other disciplines, we can build upon the legacy of ecological restoration in the region to find solutions for the challenges that we face today and in the future.

EDITORS

Christian Lenhart (Introduction, Chapters 2, 5, 9, and 10) is a research professor in the Department of Bioproducts and Biosystems Engineering at the University of Minnesota and a restoration scientist for The Nature Conservancy's Minnesota-North Dakota-South Dakota Chapter. His research focuses on restoration and management of streams and wetlands, especially in agricultural watersheds. He also teaches ecological engineering design and global water sustainability classes. Originally from Defiance, Ohio, he has also been involved in wetland restoration in the Lake Erie Basin. He has degrees from Notre Dame, the University of Wisconsin–Madison, and the University of Minnesota.

Peter C. Smiley Jr. (Introduction, Chapters 1, 2, 9, and 10) is a research ecologist with the USDA Agricultural Research Service in Columbus, Ohio. He was one of the cofounders of the Midwest–Great Lakes Chapter of the Society for Ecological Restoration (SER) and has served on the Chapter Board of Directors since 2009. In 2011 he received the John Rieger Award from SER for his contributions to establishing the Midwest–Great Lakes Chapter. He also serves as an adjunct faculty member at Indiana University–Purdue University Fort Wayne and the Ohio State University. His research specialties are restoration ecology and community ecology, and his research program focuses on providing science-based information to guide the restoration of agricultural headwater streams.

CONTRIBUTORS

Luther Aadland (Chapter 5) is a river scientist for the Minnesota Department of Natural Resources. He has conducted river research in the areas of fluvial geomorphology, river ecology, and habitat requirements of native fish and mussels for over thirty-five years. He has designed numerous projects around the United States to restore river morphology, habitat, and connectivity.

David P. Benson (Chapter 1) is a professor of biology and the science director of the Marian University Nina Mason Pulliam EcoLab, a seventy-acre natural area originally designed by Jens Jensen in 1912. He and his students are interested in the effects of restoration practices on bird and plant communities. He is the author of *Glacier Is for the Birds: A Trail Guide to the Birds of Glacier National Park.*

Andrew F. Casper (Chapter 4) is a research ecologist and field station director with the Illinois Natural History Survey. He has two decades of field experience, with a background in both freshwater ecology and coastal oceanography. A specialist in river science, his research and collaborations have spanned the continent and are often interdisciplinary, encompassing food web studies, watershed modeling, fish habitat assessment, and delta-coastal oceanography.

Hua Chen (Chapters 4 and 7) is an associate professor of biology at the University of Illinois Springfield and a guest professor at Jiyang College of Zhejiang A & F University (China). He teaches global change ecology, restoration ecology, modeling biological systems, and environmental biology as well as a scientific writing course. His research interests focus on how ecosystem restoration, climate change, and management practices influence the carbon dynamics of forests and restored wetlands.

Joe DiMisa (Chapter 6) is an instructor of environmental management at the Air Force Institute of Technology's Civil Engineer School. He joined the faculty following a thirty-year career in wetland creation and restoration, stream restoration, forest conservation, and low-impact development planning. He is a nationally recognized professional with a passion for communicating environmental topics to youth and adults.

Steve Glass (Chapter 3) is a practicing restoration ecologist in Wisconsin. He was land care manager at the University of Wisconsin–Madison Arboretum for more than twenty-five years and is co-author of the Island Press textbook *Introduction to Restoration Ecology*. He studies the history of restoration ecology and the factors that contribute to the success of long-term projects. He also uses photography to document the process of restoration ecology projects and the people who manage them.

Heath M. Hagy (Chapter 4) is a wildlife biologist with the U.S. Fish and Wildlife Service and is stationed in the Southeast Region. Previously and during development of this chapter, he served as the director of the Illinois

Natural History Survey's Forbes Biological Station and Bellrose Waterfowl Research Center at the University of Illinois. His background includes waterfowl ecology and management, avian foraging ecology, and wetland restoration and management.

John A. Harrington (Chapter 1) is professor and former chair of the Department of Landscape Architecture at the University of Wisconsin–Madison, where he has taught and conducted research in conservation planning, ecological restoration, and urban ecology for thirty-two years. His particular interest is the restoration and management of grasslands and savannas in the upper Midwest and the ecosystem services they provide. He is co-author (with Evelyn Howell and Steve Glass) of the Island Press textbook *Introduction to Restoration Ecology*.

Neil Haugerud (Chapter 5) is a river ecologist with the Minnesota Department of Natural Resources River Ecology Unit, where he has worked since 2006. He earned his master's degree in wildlife and fisheries sciences from South Dakota State University, where he studied macroinvertebrates in floodplain wetlands. He is experienced in stream biological monitoring, macroinvertebrate identification, restoration reference site selection, and evaluating water quality. His work focuses on project monitoring, river restorations, and geomorphological data analysis and management.

Constance Hausman (Chapter 6) is the plant and restoration ecologist in the Division of Natural Resources with Cleveland Metroparks, where she has worked since 2010. She implements ecosystem assessments and develops habitat recovery and restoration plans. Her current work focuses on understanding landscape-scale changes that occur due to climate change, habitat destruction, invasive species, and forest pests and pathogens.

Michael J. Lemke (Chapter 4) is a full professor in the Department of Biology at the University of Illinois Springfield, where he has instructed classes in microbiology, restoration ecology, and aquatic sciences for nearly twenty years. His research program focuses on the ecology of microorganisms of the middle reach of the Illinois River and several large river systems in Brazil. He embraces diverse professional and student collaborations.

Jen Lyndall (Chapter 6) is the certification program coordinator for the Society of Ecological Restoration (SER). In that role, she manages the

Certified Ecological Restoration Practitioner Program. Prior to SER, she worked for the National Oceanic and Atmospheric Administration as a natural resource trustee representative on Natural Resource Damage Assessments. She also worked as a consultant at Environ and ERM, with a specialization in the restoration of contaminated sediment sites and urban restoration projects.

Dan Shaw (Chapter 8) is the senior ecologist and vegetation specialist with the Minnesota Board of Water and Soil Resources. His work focuses on conservation partnerships, stormwater practices, plant community restoration, landscape resiliency, pollinator habitat, and invasive species control programs. He has also taught courses about Minnesota flora and ecological restoration planning and management at the University of Minnesota over the last fifteen years, and he has authored several publications on wetlands, stormwater design, and restoration planning.

John Shuey (Chapters 2 and 7) is the director of conservation science for the Indiana Chapter of The Nature Conservancy. In that role, he leads the chapter's efforts to develop and implement climate change adaptation and resilience strategies on over 35,000 acres of critical conservation land, including over 8,000 acres that have been restored to enhance ecological viability. He is currently engaged in efforts to assess restoration performance as a tool to reduce threats to biodiversity.

Daryl Smith (Chapter 3) is a professor emeritus of biology and science education at the University of Northern Iowa. Founder and former director of the Tallgrass Prairie Center, he has been involved in prairie preservation, management, and restoration for more than forty-five years. His former students are active in prairie restoration, management and research, secondary and college teaching, and natural history interpretation. He was executive director and producer of the documentary film *America's Lost Landscape: The Tallgrass Prairie*, which received the Pare Lorentz Award from the International Documentary Association.

REFERENCES

Aadland, L.P. 2010. *Reconnecting Rivers: Natural Channel Design in Dam Removal and Fish Passage*. Saint Paul: Minnesota Department of Natural Resources.

———. 2015. *Barrier Effects on Native Fishes of Minnesota*. Saint Paul: Minnesota Department of Natural Resources.

ACOE. 1988. *Reconnaissance Report, Wild Rice-Marsh Rivers*. Saint Paul: U.S. Army Corps of Engineers, Saint Paul District.

———. 2004. *Cross Sections for the Wild Rice River Feasibility Study*. Fargo, ND: Houston Engineering, Inc.

———. 2006. *Cross Sections for the Wild Rice River Feasibility Study 2006 Sedimentation Study Wild Rice and Marsh Creek*. Fargo, ND: Houston Engineering, Inc.

Adams, L. 1984. "Small Mammal Use of Interstate Highway Median Strip." *Journal of Applied Ecology* 21:175–78.

Ahiablame, L. M., I. Chaubey, D. R. Smith, and B. A. Engel. 2011. "Effect of Tile Effluent on Nutrient Concentration and Retention Efficiency in Agricultural Drainage Ditches." *Agricultural Water Management* 98:1271–79.

Aldrich, P., G. Parker, J. Romero-Severson, and C. Michler. 2005. "Confirmation of Oak Recruitment Failure in Indiana Old-Growth Forest: 75 Years of Data." *Forest Science* 51:406–16.

Alexander, G. G., and J. D. Allan. 2006. "Stream Restoration in the Upper Midwest, U.S.A." *Restoration Ecology* 14:595–604.

———. 2007. "Ecological Success in Stream Restoration: Case Studies from the Midwestern United States." *Environmental Management* 40:245–55.

Alexander, R. B., R. A. Smith, and G. E. Schwarz. 2000. "Effect of Stream Channel Size on Delivery of Nitrogen to the Gulf of Mexico." *Nature* 403:758–61.

Alexander, S., J. Aronson, A. Clewell, K. Keenleyside, E. Higgs, D. Martinez, C. Murcia, and C. Nelson. 2011. "Re-establishing an Ecologically Healthy Relationship between Nature and Culture: The Mission and Vision of the Society for Ecological Restoration." In *Contribution of Ecosystem Restoration to the Objectives of the CBD and a Healthy Planet for All People: Abstracts of Posters Presented at the 15th Meeting of the Subsidiary Body on Scientific, Technical and Technological Advice of the Convention on Biological Diversity*, 11–14. Montreal, CA: Secretariat of the Convention on Biological Diversity.

Anderson, J., N. Baratono, A. Streitz, J. Magner, and E. S. Verry. 2006. *Effect of Historical Logging on Geomorphology, Hydrology, and Water Quality in the Little Fork River Watershed*. Saint Paul: Minnesota Pollution Control Agency.

Anderson, W. P., D. Wall, and J. L. Olson. 2016. "Minnesota Nutrient Reduction Strategy." In *2016 10th International Drainage Symposium Conference, September 6–9, 2016*, 1–9, Minneapolis: American Society of Agricultural and Biological Engineers.

Annear, T., I. Chisholm, H. Beecher, A. Locke, P. Aarrestad, C. Coomer, C. Estes, et al. 2004. *Instream Flows for Riverine Resource Stewardship*. Cheyenne, WY: Instream Flow Council.

Anonymous. 1946. "Sportscaster." *Marian College Phoenix* (Indianapolis, IN): 4.

———. 1956. "Campus Conservation Corps Uncovers Old Trails: Plants Shrubs, Flowers to Beautify Grounds." *Marian College Phoenix* (Indianapolis, IN): 3.

———. 1959. "Board Appropriates Funds for Painting, Lights at Lakeside." *Marian College Phoenix* (Indianapolis, IN): 3.

———. 1972. "Marian Lake Area Becomes Wetland Laboratory." *Marian College Phoenix* (Indianapolis, IN): 2.

———. 1978. "Wetlands Combines Beauty, Ecological Study." *Marian College Phoenix* (Indianapolis, IN): 3.

Apfelbaum, S. I., and A. Haney. 2010. *Restoring Ecological Health to Your Land*. Washington, DC: Island Press.

Araujo, M. B., and M. New. 2007. "Ensemble Forecasting of Species Distributions." *Trends in Ecology and Evolution* 22:42–47.

Arbuckle, K., J. L. Pease, and R. Christoffel. 2004. *Managing Iowa Habitats: Restoring Iowa Streams*. Ames: Iowa State University.

Argabright, M. S., R. G. Cronshey, J. D. Helms, G. A. Pavelis, and H. R. Sinclair Jr. 1996. "Historical Changes in Soil Erosion, 1930–1992: The Northern Mississippi Valley Loess Hills." *Historical Notes* no. 5. Washington, DC: USDA Natural Resources Conservation Service.

Ashworth, W. 1986. *The Late Great Lakes*. New York: Alfred A. Knopf.

Bajer, P. G., G. Sullivan, and P. W. Sorensen. 2009. "Effects of a Rapidly Increasing Population of Common Carp on Vegetative Cover and Waterfowl in a Recently Restored Midwestern Shallow Lake." *Hydrobiologia* 632:235–45.

Baker, H. G. 1974. "The Evolution of Weeds." *Annual Review of Ecology and Systematics* 5:1–24.

———. 1986. "Patterns of Plant Invasion in North America." In *Ecology of*

Biological Invasions of North America and Hawaii, edited by H. A. Mooney and J. A. Drake, 44–57. New York: Springer-Verlag.

Ballantine, K., and R. Schneider. 2009. "Fifty-Five Years of Soil Development in Restored Freshwater Depressional Wetlands." *Ecological Applications* 19:1467–80.

Bavarian State Ministry for Regional Development and the Environment (BSMRDE). 1997. *Water Management in Bavaria: Rivers, Meadows, Valleys Preserve and Develop*. Bavaria: Bavarian State Ministry for Regional Development and the Environment.

Bednarek, A. T. 2001. "Undamming Rivers: A Review of the Ecological Impacts of Dam Removal." *Environmental Management* 27:803–14.

Bellmore, J. R., J. J. Duda, L. S. Craig, S. L. Greene, C. E. Torgersen, M. J. Collins, and K. Vittum. 2017. "Status and Trends of Dam Removal Research in the United States." *Wires Water* 4:E1164.

Bellrose, F. C., S. P. Havera, F. L. Paveglio Jr., and D. W. Steffeck. 1983. "The Fate of Lakes in the Illinois River Valley." *Illinois Natural History Survey Biological Notes* 119. Champaign, IL: Illinois Natural History Survey.

Bellrose, F. C., F. L. Paveglio Jr., and D. W. Steffeck. 1979. "Waterfowl Populations and the Changing Environment of the Illinois River Valley." *Illinois Natural History Survey Bulletin* 32:1–54.

BenDor, T., T. W. Lester, A. Livengood, A. Davis, and L. Yonavjak. 2015. "Estimating the Size and Impact of the Ecological Restoration Economy." *PLOS ONE* 10: E0128339. doi:10.1371/journal.pone.0128339.

Benson, D. P. 2004. "Jensen Designed Property as a Case Study of the Long-Term Efficacy of Ecological Restoration." *Proceedings of the 16th International Conference of the Society for Ecological Restoration*. Tuscon, AZ: Society for Ecological Restoration.

Berliner, R., and J. G. Torrey. 1989. "Studies on Mycorrhizal Associations in Harvard Forest, Massachusetts." *Canadian Journal of Botany* 67:2245–51.

Bernal, B., and W. J. Mitsch. 2012. "Comparing Carbon Sequestration in Temperate Freshwater Wetland Communities." *Global Change Biology* 18:1636–47.

Bernhardt, E. S., and M. A. Palmer. 2011. "River Restoration: The Fuzzy Logic of Repairing Reaches to Reverse Catchment Scale Degradation." *Ecological Applications* 21:1926–31.

Bernhardt, E. S., M. A. Palmer, J. D. Allan, G. Alexander, K. Barnas, S. Brooks, J. Carr, et al. 2005. "Synthesizing U.S. River Restoration Efforts." *Science*

308:636–637.

Bernhardt, E. S., E. B. Sudduth, M. A. Palmer, J. D. Allan, J. L. Meyer, G. Alexander, J. Follastad-Shah, et al. 2007. "Restoring Rivers One Reach at a Time: Results from a Survey of U.S. River Restoration Practitioners." *Restoration Ecology* 15:482–93.

Biohabitats. 2014. *Acacia Reservation: Ecological Restoration Master Plan.* Cleveland, OH: Cleveland Metroparks.

Biondini, M. 2007. "Plant Diversity, Production, Stability, and Susceptibility to Invasion in Restored Northern Tall Grass Prairies (United States)." *Restoration Ecology* 15:77–87.

Blann, K. L., J. L. Anderson, G. R. Sands, and B. Vondracek. 2009. "Effects of Agricultural Drainage on Aquatic Ecosystems: A Review." *Critical Reviews in Environmental Science and Technology* 39:909–1001.

Blewett, T. 1981. "An Ordination Study of Plant Species Ecology in the Arboretum Prairies." PhD diss., University of Wisconsin–Madison.

Blodgett, K. D. 2013. *A Summary of the Great Flood of 2013 at The Nature Conservancy's Emiquon Preserve.* Lewistown, IL: The Nature Conservancy.

Bohlen, D. A. 1946. Letter to Reverend Sister Mary Cephas, Marian College. Mother Theresa Hackelmeier Memorial Library Archives. Marian University, Indianapolis, IN.

Botkin, D. B. 1990. *Disconcordant Harmonies: A New Ecology for the Twenty-First Century.* New York: Oxford University Press.

Bouldin, J. L., J. L. Farris, M. T. Moore, S. Smith Jr., and C. M. Cooper. 2005. "Evaluated Fate and Effects of Atrazine and Lambda-Cyhalothrin in Vegetated and Unvegetated Microcosms." *Environmental Toxicology* 20:487–98.

Bowles, M., S. Apfelbaum, A. Haney, S. Lehnhardt, and T. Post. 2011. "Canopy Cover and Groundlayer Vegetation Dynamics in a Fire Managed Eastern Sand Savanna." *Forest Ecology and Management* 262:1972–82.

Bowyer, M. W., J. D. Stafford, A. P. Yetter, C. S. Hine, M. M. Horath, and S. P. Havera. 2005. "Moist-Soil Plant Seed Production for Waterfowl at Chautauqua National Wildlife Refuge, Illinois." *The American Midland Naturalist* 154:331–41.

Bradshaw, A. D. 1987. "Restoration: An Acid Test for Ecology." In *Restoration Ecology: A Synthetic Approach to Ecological Research*, edited by W. R. Jordan III., M. E. Gilpin, and J. D. Aber, 23–29. Cambridge: Cambridge University Press.

Brandt, J., K. Henderson, and J. Uthe. 2011. *Integrated Roadside Vegetation Management: Technical Manual*. Cedar Falls, IA: Tallgrass Prairie Center, University of Northern Iowa.

Brandt, L. H., L. Iverson, F. Thompson, P. Butler, S. Handler, M. Janowiak, P. Shannon, et al. 2014. *Central Hardwoods Ecosystem Vulnerability Assessment and Synthesis: A Report from the Central Hardwoods Climate Change Response Framework Project*. General Technical Report NRS-124. Newtown Square, PA: USDA Forest Service Northern Research Station.

Briddell, B. 2012. "Carbon and Nitrogen Storage in Natural Illinois Wetlands: Comparing Marshes and Sedge Meadows." Master's thesis, University of Illinois Springfield.

Broadman, R., L. Parrish, H. Kraus, and S. Cortwright. 2006. "Amphibian Biodiversity Recovery in a Large-Scale Ecosystem Restoration." *Herpetological Conservation and Biology* 1:101–8.

Broennimann, O., P. Mraz, B. Petitpierre, A. Guisan, and H. Muller-Scharer. 2014. "Contrasting Spatio-Temporal Climatic Niche Dynamics during the Eastern and Western Invasions of Spotted Knapweed in North America." *Journal of Biogeography* 41:1126–36.

Brookes, A. 1988. *Channelized Rivers: Perspectives for Environmental Management*. Chichester, UK: Wiley.

———. 1989. "Alternative Channelization Procedures." In *Alternatives in Regulated River Management*, edited by J. A. Gore and G. E. Petts, 139–62. Boca Raton, FL: CRC Press.

Brooks, T. C. 2005. *The Nature of Wetland Restoration: The LaGrange Wetland Mitigation Bank Site, Brown County, Illinois*. Springfield: Illinois Department of Transportation.

Broughton, J., and S. Apfelbaum. 1999. "Using Ecological Systems for Alternative Stormwater Management." *Land and Water Magazine* (September/October): 10–14.

Brundrett, M. C. 1991. "Mycorrhizas in Natural Ecosystems." *Advances in Ecological Research* 21:171–313.

———. 2009. "Mycorrhizal Associations and Other Means of Nutrition of Vascular Plants: Understanding the Global Diversity of Host Plants by Resolving Conflicting Information and Developing Reliable Means of Diagnosis." *Plant and Soil* 320:37–77.

Buerger, A., K. Howe, E. Jacquart, M. Chandler, T. Culley, C. Evans, K. Kearns, et al. 2016. "Risk Assessments for Invasive Plants: A Midwestern U.S.

Comparison." *Invasive Plant Science and Management* 9:41–54.

Bunn, S. E., and A. H. Arthington. 2002. "Basic Principles and Ecological Consequences of Altered Flow Regimes for Aquatic Biodiversity." *Environmental Management* 30:492–507.

Burgess, R. L. 1976. *The Ecological Society of America: Historical Data and Some Preliminary Analyses*. Oak Ridge, TN: Oak Ridge National Laboratory.

Buss, C. 2014. "Storage of Soil Organic Matter (SOC) and Total Nitrogen (TN) in Restored Wetlands from Croplands of Illinois: A Chronosequence Approach." Master's thesis, University of Illinois Springfield.

Cabin, R. J. 2011. *Intelligent Tinkering: Bridging the Gap between Science and Practice*. Washington, DC: Island Press.

Cadotte, M. W., and J. Lovett-Doust. 2001. "Ecological and Taxonomic Differences between Native and Introduced Plants of Southwestern Ontario." *Ecoscience* 8:230–38.

Camp, M., and L. Best. 1994. "Nest Density and Nesting Success of Birds in Roadsides Adjacent to Rowcrop Fields." *The American Midland Naturalist* 131:347–58.

Carpenter, S. R., and J. F. Kitchell. 1996. *The Trophic Cascade in Lakes*. Cambridge: Cambridge University Press.

Carpenter, S. R., J. F. Kitchell, and J. R. Hogson. 1985. "Trophic Interactions and Lake Productivity." *Bioscience* 35:634–39.

Carpenter, S. R., D. Ludwig, and W. A. Brock. 1999. "Management of Eutrophication for Lakes Subjected to Potentially Irreversible Change." *Ecological Applications* 9:751–71.

Cassidy, V. M. 2007. *Henry Chandler Cowles – Pioneer Ecologist*. Chicago: Kedziesigel Press.

CBD. 2011. *Strategic Plan for Biodiversity 2011–2020, Including the Aichi Biodiversity Targets*. Montreal, CA: Secretariat of the Convention on Biological Diversity. Accessed June 30, 2017. https://www.cbd.int/sp/targets/.

Chazdon, R. 2008. "Beyond Deforestation: Restoring Forests and Ecosystem Services on Degraded Lands." *Science* 320:1458–60.

Chen, H., S. Popovich, A. McEuen, and B. Briddell. 2017. "Carbon and Nitrogen Storage of a Restored Wetland at Illinois' Emiquon Preserve: Potential for Carbon Sequestration." *Hydrobiologia* 804:139-50.

Chenoweth, J., S. A. Acker, and M. L. McHenry. 2011. *Revegetation and Restoration Plan for Lake Mills and Lake Aldwell*. Port Angeles, WA: Olympic

National Park and the Lower Elwha Klallam Tribe.

Choi, Y. D., V. M. Temperton, E. B. Allen, A. P. Grootjans, R. J. Hobbs, M. A. Naeth, and K. Torok. 2008. "Ecological Restoration for Future Sustainability in a Changing Environment." *Ecoscience* 15:53–64.

Christiansen, P., and D. Lyons. 1975. *Research Report on Roadside Vegetation Management.* Ames: Iowa Department of Transportation.

Clark, O. R. 1937. "Interception of Rainfall by Herbaceous Vegetation." *Science* 86:591–92.

Classen, R., F. Carriazo, J. C. Cooper, D. Hellerstein, and K. Udea. 2011. *Grassland to Cropland Conversion in the Northern Plains: The Role of Crop Insurance, Commodity, and Disaster Programs.* Economic Research Report #120. Washington, DC: USDA Economic Research Service.

Clements, F. E. 1916. *Plant Succession: An Analysis of the Development of Vegetation.* Publication 242. Washington, DC: Carnegie Institution of Washington.

———. 1936. "Nature and Structure of the Climax." *Journal of Ecology* 24:252–84.

Clewell, A. 2009. "Intent of Ecological Restoration, Its Circumscription, and Its Standards." *Ecological Restoration* 27:5–7.

Colvin, M., C. L. Pierce, T. W. Stewart, and S. E. Grummer. 2012. "Strategies to Control a Common Carp Population by Pulsed Commercial Harvest." *North American Journal of Fisheries Management* 32:1251–64.

Connor, J. E., J. J. Duda, and G. E. Grant. 2015. "1000 Dams Down and Counting." *Science* 348:496–97.

Cooke, G. D. 2007. "History of Eutrophic Lake Rehabilitation in North America with Arguments for Including Social Sciences in the Paradigm." *Lake and Reservoir Management* 23:323–29.

Cooke, G. D., E. B. Welch, S. A. Peterson, and S. A. Nichols. 2005. *Restoration and Management of Lakes and Reservoirs.* Boca Raton, FL: Taylor and Francis.

Cordova, J. M., E. J. Rosi-Marshall, A. M. Yamamuroa, and G. A. Lamberti. 2007. "Quantity, Controls, and Functions of Large Woody Debris in Midwestern USA Streams." *River Research and Applications* 23:21–33.

Cornett, M., and M. White. 2013. "Forest Restoration in a Changing World: Complexity and Adaptation Examples from the Great Lakes Region of North America." In *Managing Forests as Complex Adaptive Systems: Building Resilience to the Challenge of Global Change,* edited by C. Messier, K.

Puettmann, and D. Coates, 113–32. New York: Routledge.

Cottam, G. 1987. "Community Dynamics on an Artificial Prairie." In *Restoration Ecology: A Synthetic Approach to Ecological Research*, edited by W. R. Jordan III, M. E. Gilpin, and J. D. Aber, 257–70. Cambridge: Cambridge University Press.

Cottam, J. T., and H. C. Wilson. 1966. "Community Dynamics on an Artificial Prairie." *Ecology* 47:88–96.

Craft, C., P. Megonigal, S. Broome, J. Stevenson, R. Freese, J. Cornell, L. Zheng, and J. Sacco. 2003. "The Pace of Ecosystem Development of Constructed *Spartina alterniflora* Marshes." *Ecological Applications* 13:1417–32.

Crall, W., G. J. Newman, C. S. Jarnevich, T. J. Stohlgren, D. M. Waller, and J. Graham. 2010. "Improving and Integrating Data on Invasive Species Collected by Citizen Scientists." *Biological Invasions* 12:3419–28.

Cramer, C. 1991. "Tougher Than Weeds." *The New Farm* 13:37–39.

Croker, R. A. 1991. *Pioneer Ecologist: The Life and Work of Victor Ernest Shelford 1877–1968*. Washington, DC: Smithsonian Institution Press.

———. 2001. *Stephen Forbes and the Rise of American Ecology*. Washington, DC: Smithsonian Institution Press.

Cronon, W. 2009. *Nature's Metropolis: Chicago and the Great West*. New York: W. W. Norton and Company.

Curtis, J. T. 1959. *The Vegetation of Wisconsin: An Ordination of Plant Communities*. Madison: University of Wisconsin Press.

Curtis, J. T., and G. Cottam. 1950. "Antibiotic and Autotoxic Effects in Prairie Sunflower." *Bulletin of the Torrey Botany Club* 77:187–91.

Curtis, J. T., and M. L. Partch. 1948. "Effect of Fire on the Competition between Blue Grass and Certain Prairie Plants." *The American Midland Naturalist* 39:437–43.

Dahl, T. E. 1990. *Wetlands Losses in the United States, 1780s to 1980s*. Washington, DC: U.S. Fish and Wildlife Service.

Dahl, T. E., and G. J. Allord. 1997. *Technical Aspects of Wetlands: History of Wetlands in the Conterminous United States*. Water Supply Paper 2425. Washington, DC: U.S. Geological Survey.

Daily, G. C. 1997. *Nature's Services: Societal Dependence on Natural Systems*. Washington, DC: Island Press.

Dale, V., L. Joyce, S. McNulty, R. Neilson, M. Ayres, M. Flannigan, P. Hanson, et al. 2001. "Climate Change and Forest Disturbances." *Bioscience* 51:723–34.

Delaplane, K., and D. Mayer. 2000. *Crop Pollination by Bees*. New York: CBI

Publishing.

Delong, M. D. 2005. "Upper Mississippi River Basin." In *Rivers of North America*, edited by A. C. Benke and C. E. Cushing, 327–73. Burlington, MA: Elsevier Academic Press.

Dent, C. L., G. S. Cumming, and S. R. Carpenter. 2002. "Multiple States in River and Lake Ecosystems." *Philosophical Transactions of the Royal Society* (B) 357:635–45.

Dierks, S. 2011. "Quantifying Prairie and Forest Impacts on Soil Water Holding Capacity and Infiltration." *Stormwater: Journal for Surface Water Quality Professionals*. Accessed April 19, 2017. http://foresternetwork.com/daily/water/quantifying-prairie-and-forest-impacts-on-soil-water-holding-capacity-and-infiltration/.

Dodson, K. R. 1998. "The Work of Jens Jensen at the James A. Allison Estate." Master's thesis, Ball State University.

Dopplet, B., M. Scurlock, C. Frissell, and J. Karr. 1993. *Entering the Watershed: A New Approach to Save America's River Ecosystems*. Washington, DC: Island Press.

Doyle, M. W., E. H. Stanley, C. H. Orr, A. R. Selle, S. A. Sethi, and J. M. Harbor. 2005. "Stream Ecosystem Response to Small Dam Removal: Lessons from the Heartland." *Geomorphology* 71:227–44.

Dunlop, T. R. 1980. "The Gypsy Moth: A Study in Science and Public Policy." *Journal of Forest History* 21:116–26.

Dunne, T., and L. B. Leopold. 1978. *Water in Environmental Planning*. New York: W. H. Freeman and Company.

Ebersole, J. L., W. J. Liss, and C. A. Frissel. 1997. "Restoration of Stream Habitats in the Western United States: Restoration as a Re-expression of Habitat Capacity." *Environmental Management* 21:1–14.

EC WFC. 2006. *Euclid Creek Watershed Action Plan: Protection, Restoration, and Management for the Future*. Valley View, OH: Euclid Creek Watershed Council, Friends of Euclid Creek, and Cuyahoga Soil and Water Conservation District.

Egan, D., and E. A. Howell. 2001. *The Historical Ecology Handbook: A Restorationist's Guide to Reference Ecosystems*. Washington, DC: Island Press.

Egan, D., and W. H. Tishler. 1999. "Jens Jensen, Native Plants, and the Concept of Nordic Superiority." *Landscape Journal* 17:11–29.

Engel, R. J. 1983. *Sacred Sands: The Struggle for Community in the Indiana Dunes*. Middletown, CT: Wesleyan University Press.

Environmental Law Institute. 2004. *Making a List: Prevention Strategies for Invasive Plants in the Great Lakes States*. Washington, DC: Environmental Law Institute.

Environmental Protection Agency (EPA). 2015. *Mississippi River/Gulf of Mexico Watershed Nutrient Task Force 2015 Report to Congress*. Washington, DC: U.S. Environmental Protection Agency.

Euliss, N. H., Jr., R. A. Gleason, A. Olness, R. L. McDougal, H. R. Murkin, R. D. Robarts, R. A. Bourbonniere, and B. G. Warner. 2006. "North American Prairie Wetlands Are Important Nonforested Land-Based Carbon Storage Sites." *Science of the Total Environment* 361:179–88.

Falk, D. A., M. A. Palmer, and J. B. Zedler. 2006. *Foundations of Restoration Ecology*. Washington, DC: Island Press.

Federal Emergency Management Agency (FEMA). 2006. *Geotextiles in Embankment Dams*. Denver: U.S. Federal Emergency Management Agency.

Fisichelli, N., L. Frelich, and P. Reich. 2014. "Temperate Tree Expansion into Adjacent Boreal Forest Patches Facilitated by Warmer Temperatures." *Ecography* 37:152–61.

Flinn, K. M., J. Bechhofer, and M. Malcolm. 2014. "Little Impact of the Invasive Shrub Japanese Barberry (*Berberis thunbergii* DC) on Forest Understory Plant Communities." *The Journal of the Torrey Botanical Society* 141:217–24.

Forbes, S. A. 1887. "The Lake as a Microcosm." In *Bulletin of the Peoria Scientific Association*, 77–87. Peoria, IL: Edward Hine Jr. Company.

Forman, R., T. Sperling, J. Bissonette, A. Clevenger, C. Cutshall, V. Dale, L. Fahrig, et al. 2003. *Road Ecology: Science and Solutions*. Washington, DC: Island Press.

Fralish, J. S., R. P. McIntosh, and O. L. Loucks. 1993. *John T. Curtis: Fifty Years of Wisconsin Plant Ecology*. Madison: Wisconsin Academy of Sciences, Arts and Letters.

Francis, R. A., and S. P. G. Hoggart. 2008. "Waste Not, Want Not: The Need to Utilize Existing Artificial Structures for Habitat Improvement along Urban Rivers." *Restoration Ecology* 16:373–81.

Frelich, L. E., and A. R. Holdsworth. 2002. *Exotic Earthworms in Minnesota Hardwood Forests: An Investigation of Earthworm Distribution, Understory Plant Communities, and Forest Floor Dynamics in Northern Hardwood Forests*. Saint Paul: Department of Forest Resources, University of Minnesota.

Fridley, J. D., and D. F. Sax. 2014. "The Imbalance of Nature: Revisiting a Darwinian Framework for Invasion Biology." *Global Ecology and*

Biogeography 23:1157–66.

Frissell, C. A., W. J. Liss, R. K. Nawa, R. E. Gresswell, and J. L. Ebersole. 1997. "Measuring the Failure of Salmon Management." In *Pacific Salmon and Their Ecosystems: Status and Future Options*, edited by D. J. Stouder, P. A. Bisson, and R. J. Naiman, 411–44. New York: Chapman and Hall.

Frissell, C. A., and R. K. Nawa. 1992. "Incidence and Causes of Physical Failure of Artificial Habitat Structures in Streams of Western Oregon and Washington." *North American Journal of Fisheries Management* 12:182–97.

Frissell, C. A., and S. C. Ralph. 1998. "Stream and Watershed Restoration." In *River Ecology and Management: Lessons from the Pacific Coastal Ecoregion*, edited by R. J. Naiman and R. E. Bilby, 599–624. New York: Springer-Verlag.

FSA. 2011. *Monthly Conservation Reserve Program Summary – December 2011*. Washington, DC: USDA Farm Service Agency.

———. 2012. CRP *Enrollment and Rental Payments by State, 1986–2011*. Washington, DC: USDA Farm Service Agency. Accessed March 28, 2017. http://www.fsa.usda.gov/internet/fsa_file/historystate121911.xls.

Funderberg, E. 2001. "What Does Organic Matter Do in Soil?" *Ag News and Views*. Accessed April 19, 2017. https://www.noble.org/news/publications/ag-news-and-views/2001/august/what-does-organic-matter-do-in-soil/.

Galat, D. L., L. H. Fredrickson, D. D. Humburg, K. T. Bataille, J. R. Bodie, J. Dohrenwend, G. T. Gelwicks, et al. 1998. "Flooding to Restore Connectivity of Regulated, Large-River Wetlands." *Bioscience* 48:721–33.

Galatowitsch, S. M. 2012. *Ecological Restoration*. Sunderland, MA: Sinauer Associates.

Galatowitsch, S., L. Frelich, and L. Phillips-Mao. 2009. "Regional Climate Change Adaptation Strategies for Biodiversity Conservation in a Midcontinental Region of North America." *Biological Conservation* 142:2012–22.

Galatowitsch, S., and A. van der Valk. 1994. *Restoration of Prairie Pothole Wetlands: An Ecological Approach*. Ames: Iowa State University Press.

Gannon, J. J., T. L. Shaffer, and C. T. Moore. 2013. *Native Prairie Adaptive Management: A Multi Region Adaptive Approach to Invasive Plant Management on Fish and Wildlife Service Owned Native Prairies*. U.S. Geological Survey Open File Report 2013–1279. Washington, DC: United States Geological Survey.

Garvey, J. E., J. C. Chick, M. W. Eichholz, G. Conover, and R. C. Brooks. 2007. *Swan Lake Habitat Rehabilitation and Enhancement Project: Post-Project*

Monitoring of Water Quality, Sedimentation, Vegetation, Invertebrates, Fish Communities, Fish Movement, and Waterbirds. Final Report. Carbondale: Southern Illinois University.

Gee, M. A. 1952. *Fish Stream Improvement Handbook.* Washington, DC: U.S. Forest Service.

Ghane, E., N. R. Fausey, V. S. Shedekar, H. P. Piepho, Y. Shang, and L. C. Brown. 2012. "Crop Yield Evaluation under Controlled Drainage in Ohio, United States." *Journal of Soil and Water Conservation* 67:465–73.

Gleason, H. A. 1917. "The Structure and Development of the Plant Association." *Bulletin of the Torrey Botanical Club* 43:463–81.

Gobster, P. H. 2000. "Restoring Nature: Human Actions, Interactions, and Reactions." In *Restoring Nature: Perspectives from the Social Sciences and Humanities*, edited by P. H. Gobster and R. B. Hull, 1–19. Washington, DC: Island Press.

———. 2010. "Introduction: Urban Ecological Restoration." *Nature and Culture* 5:227–30.

Gorman, O. T., and J. R. Karr. 1978. "Habitat Structure and Stream Fish Communities." *Ecology* 59:507–15.

Grams, P. E., J. C. Schmidt, S. A. Wright, D. J. Topping, T. S. Melis, and D. M. Rubin. 2015. "Building Sandbars in the Grand Canyon." *Eos* 96:12–16.

Green, E. K., and S. M. Galatowitsch. 2002. "Effects of *Phalaris arundinacea* and Nitrate-N Addition on the Establishment of Wetland Plant Communities." *Journal of Applied Ecology* 39:134–44.

Greene, H. C., and J. T. Curtis. 1950. "Germination Studies of Wisconsin Prairie Plants." *The American Midland Naturalist* 43:186–94.

Gregory, S. V., K. L. Boyer, and A. M. Gurnell. 2003. *The Ecology and Management of Wood in World Rivers.* Symposium 37. Bethesda, MD: American Fisheries Society.

Grese, R. E. 1992. *Jens Jensen: Maker of Natural Parks and Gardens.* Baltimore, MD: Johns Hopkins University Press.

———. 1995. "The Prairie Gardens of O. C. Simonds and Jens Jensen." In *Regional Garden Design in the United States*, edited by T. O'Malley and M. Treib, 99–123. Washington, DC: Dumbarton Oaks Research Library and Collection.

———. 2011. *The Native Landscape Reader.* Amherst: University of Massachusetts Press.

Griffiths, N. A., J. L. Tank, S. S. Roley, and M. L. Stephen. 2012. "Decomposition of Maize Leaves and Grasses in Restored Agricultural Streams." *Freshwater Science* 31:848–64.

Grumbine, R. E. 1994. *Environmental Policy and Biodiversity*. Washington, DC: Island Press.

Gustafson, E., and B. Sturtevant. 2013. "Modeling Forest Mortality Caused by Drought Stress: Implications for Climate Change." *Ecosystems* 16:60–74.

Hagy, H. M., C. S. Hine, M. M. Horath, A. P. Yetter, R. V. Smith, and J. D. Stafford. 2017. "Waterbird Response Indicates Floodplain Wetland Restoration." *Hydrobiologia* 804:119-37.

Haney, A., M. Bowles, A. Apfelbaum, E. Lain, and T. Post. 2008. "Gradient Analysis of an Eastern Sand Savanna's Woody Vegetation, and Its Long-Term Responses to Restored Fire Processes." *Forest Ecology and Management* 266:1560–71.

Harris, J. A., R. J. Hobbs, E. Higgs, and J. Aronson. 2006. "Ecological Restoration and Global Climate Change." *Restoration Ecology* 14:170–76.

Harwell, M. A., J. H. Gentile, A. Bartuska, C. C. Harwell, V. Meyers, J. Obeysekera, J. C. Ogden, and S. C. Tosini. 1999b. "A Science-based Strategy for Ecological Restoration in South Florida." *Urban Ecosystems* 3:201–222.

Harwell, M. A., V. Myers, T. Young, A. Bartuska, N. Gassman, J. H. Gentile, C. C. Harwell, et al. 1999a. "A Framework for an Ecosystem Integrity Report Card." *Bioscience* 49:543–56.

Hasler, A. D. 1947. "Eutrophication of Lakes by Domestic Drainage." *Ecology* 28:383–95.

———. 1964. "Experimental Limnology." *Bioscience* 14:36–38.

Havera, S. P., K. E. Roat, and L. M. Anderson. 2003. *The Thompson Lake/ Emiquon Story: The Biology, Drainage, and Restoration of an Illinois River Bottomland Lake*. Urbana: Illinois Natural History Survey.

Heidkamp, C. 1952. "Mr. Clemens Rounds 32 Years of Service in Campus Maintenance, Development." *Marian College Phoenix* (Indianapolis, IN): 3.

Heinz Center. 2002. *Dam Removal: Science and Decision Making*. Washington, DC: The Heinz Center.

Heller, N., and E. Zavaleta. 2008. "Biodiversity Management in the Face of Climate Change: A Review of 22 Years of Recommendations." *Biological Conservation* 142:14–32.

Helzer, C. 2009. *The Ecology and Management of Prairies in the Central United States*. Iowa City: University of Iowa Press.

———. 2012. "The Right Metaphor for Prairie Restoration." *The Prairie*

Ecologist. Accessed April 9, 2017. https://prairieecologist.com/2012/11/06/the-right-metaphor-for-prairie-restoration/.

Henry, C., and M. Watland. 2009. *Becker County Cooperative Weed Management Plan: Pulling Together in Becker County.* Detroit Lakes, MN: Partnership Committee.

Herms, D. A., and D. G. McCullough. 2013. "Emerald Ash Borer Invasion of North America: History, Biology, Ecology, Impacts, and Management." *Annual Review of Entomology* 59:13–30.

Higgs, E. 2003. *Nature by Design: People, Natural Processes, and Ecological Restoration.* Cambridge, MA: MIT Press.

Hilderbrand, R., A. Watts, and A. Randle. 2005. "The Myths of Restoration Ecology." *Ecology and Society* 10:19. Accessed February 10, 2017. http://www.ecologyandsociety.org/vol10/iss1/art19/.

Hine, C. S., H. M. Hagy, M. M. Horath, A. P. Yetter, R. V. Smith, and J. D. Stafford. 2017. "Response of Aquatic Vegetation Communities and Other Wetland Cover Types to Floodplain Restoration at Emiquon Preserve." *Hydrobiologia* 804: 59–71.

Hobbs, R. J. 2007. "Setting Effective and Realistic Restoration Goals: Key Directions for Research." *Restoration Ecology* 15:354–57.

Hobbs, R. J., and J. A. Harris. 2001. "Restoration Ecology: Repairing the Earth's Ecosystems in the New Millennium." *Restoration Ecology* 9:239–46.

Hobbs, R. J., E. Higgs, and J. A. Harris. 2009. "Novel Ecosystems: Implications for Conservation and Restoration." *Trends in Ecology and Evolution* 24:599–605.

Hobbs, R. J., and K. N. Suding. 2009. *New Models for Ecosystem Dynamics and Restoration.* Washington, DC: Island Press.

Hobbs, W. O., J. M. Ramstack, T. LaFrancois, K. D. Zimmer, K. M. Theissen, M. B. Edlund, N. Micheluti, et al. 2012. "A 200-Year Perspective on Alternative Stable State Theory and Lake Management from a Biomanipulated Shallow Lake." *Ecological Applications* 22:1483–96.

Homoya, M. 1997. "Land of the Cliff Dwellers: The Shawnee Hills Natural Region." In *The Natural Heritage of Indiana,* edited by M. Jackson, 172–76. Bloomington: Indiana University Press.

Homoya, M., and H. Huffman. 1997. "Sinks, Slopes, and a Stony Disposition: The Highland Rim Natural Region." In *The Natural Heritage of Indiana,* edited by M. Jackson, 167–71. Bloomington: Indiana University Press.

Homoya, M. A., D. B. Abrell, J. R. Aldrich, and W. P. Thomas. 1985. "The Natural

Regions of Indiana." *Proceedings of the Indiana Academy of Sciences* 94:245–68.

Howell, E., A. Harrington, and S. Glass. 2012. *Introduction to Restoration Ecology.* Washington, DC: Island Press.

Howell, E., and F. Stearns. 1993. "The Preservation, Management, and Restoration of Wisconsin Plant Communities: The Influence of John Curtis and His Students." In *John T. Curtis: Fifty Years of Wisconsin Plant Ecology,* edited by J. S. Fralish, R. P. McIntosh, and O. L. Loucks, 7–66. Madison: Wisconsin Academy of Sciences, Arts, and Letters.

Ingenloff, K., A. Lira-Noriega, N. Barve, V. Barve, H. Owens, C. Hensz, A. Townsend Peterson, and J. Soberón. 2011. *Risk Analysis of the Invasive Potential of 6 Species in Iowa Utilizing Ecological Niche Modeling to Assess Climatic Suitability in 2050 and 2090.* Lawrence: Ecological Niche Modeling Working Group, University of Kansas.

Isenhart, T. M., R. C. Schultz, and J. P. Colletti. 1997. "Watershed Restoration and Agricultural Practices in the Midwest: Bear Creek of Iowa." In *Watershed Restoration: Principles and Practices,* edited by J. E. Williams, C. A. Wood, and M. P. Dombeck, 318–34. Bethesda, MD: American Fisheries Society.

Jackson, C. R., and C. M. Pringle. 2010. "Ecological Benefits of Reduced Hydrologic Connectivity in Intensively Developed Landscapes." *Bioscience* 60:37–46.

Jackson, D. L., and L. L. Jackson. 2002. *The Farm as Natural Habitat.* Washington, DC: Island Press.

Janssen, J. R. 2008. "Environmental and Management Influences on Fish and Invertebrate Communities in Agricultural Headwater Systems." Master's thesis, University of Michigan.

Jarchow, M. E., and M. Liebman. 2011. *Incorporating Prairies into Multifunctional Landscapes: Establishing and Managing Prairies for Enhanced Environmental Quality, Livestock Grazing and Hay Production, Bioenergy Production, and Carbon Sequestration.* PMR 1007. Ames: Iowa State University Extension.

Jaynes, D. B., and T. M. Isenhart. 2014. "Reconnecting Tile Drainage to Riparian Buffer Hydrology for Enhanced Nitrate Removal." *Journal of Environmental Quality* 43:631–38.

Jensen, J. 1939. *Siftings.* Chicago: Ralph Fletcher Seymour.

Jeppesen, E., J. P. Jensen, P. Kristensen, M. Søndergaard, E. Mortensen, O. Sortkjaer, and K. Olrik. 1990. "Fish Manipulation as a Lake Restoration Tool in Shallow, Eutrophic, Temperate Lakes 2: Threshold Levels, Long-Term Stability, and Conclusions." *Hydrobiologia* 200/201:219–27.

Johnson, K. A., S. Polasky, E. Nelson, and D. Pennington. 2012. "Uncertainty in Ecosystem Services Valuation and Implications for Assessing Land Use Tradeoffs: An Agricultural Case Study in the Minnesota River Basin." *Ecological Economics* 79:71–79.

Johnson, W. D., and A. D. Hasler. 1964. "Rainbow Trout Production in Dystrophic Lakes." *Journal of Wildlife Management* 18:113–34.

Jones, H. P., and O. J. Schmitz. 2009. "Rapid Recovery of Damaged Ecosystems." *Plos ONE* 4: E5653. doi:10.1371/journal.pone.0005653.

Jordan, W. R., III. 2003. *The Sunflower Forest: Ecological Restoration and the New Communion with Nature.* Berkeley: University of California Press.

Jordan, W. R., III, M. E. Gilpin, and J. D. Aber. 1987. *Restoration Ecology: A Synthetic Approach to Ecological Research.* Cambridge: Cambridge University Press.

Jordan, W. R., III, and G. M. Lubick. 2011. *Making Nature Whole: A History of Ecological Restoration.* Washington, DC: Island Press.

Jørgensen, U. 2011. "Benefits versus Risks of Growing Biofuel Crops: The Case of *Miscanthus.*" *Current Opinion in Environmental Sustainability* 3:24–30.

Junk, W. J., P. B. Bayley, and R. E. Sparks. 1989. "The Flood Pulse Concept in River-Floodplain Systems." *Canadian Special Publications of Fisheries and Aquatic Sciences* 106:110–27.

Kane, D. D., J. D. Conroy, R. P. Richards, D. B. Baker, and D. A. Culver. 2014. "Re-eutrophication of Lake Erie: Correlations between Tributary Nutrient Loads and Phytoplankton Biomass." *Journal of Great Lakes Research* 40:496–501.

Karr, J. R. 1981. "Assessment of Biotic Integrity Using Fish Communities." *Fisheries* 6:21–27.

Karr, J. R., L. A. Toth, and D. R. Dudley. 1985. "Fish Communities of Midwestern Rivers: A History of Degradation." *Bioscience* 35:90–95.

Kingsbury, B. A., and J. Gibson. 2002. *Habitat Management Guidelines for Amphibians and Reptiles of the Midwest.* Technical Publication HMG-1. Fort Wayne, IN: Partners in Amphibian and Reptile Conservation.

Klaus, V. H. 2013. "Urban Grassland Restoration: A Neglected Opportunity for Biodiversity Conservation." *Restoration Ecology* 21:665–69.

Kline, B. 2011. *First along the River: A Brief History of the U.S. Environmental Movement.* Lanham, MD: Rowman and Littlefield.

Kline, V. M. 1985. "Response of Sweet Clover (*Meliotus alba*) and Associated Prairie Vegetation of Seven Experimental Burning and Mowing Regimes."

In *Proceedings of the 9th North American Prairie Conference*, edited by G. K. Clambey and R. H. Pemble, 149–52. Fargo, IA: Tri-College Press.

———. 1992. *The Long-Range Management Plan for Arboretum Ecological Communities*. Madison: University of Wisconsin Press.

———. 1993. "John Curtis and the University of Wisconsin Arboretum." In *John T. Curtis: Fifty Years of Wisconsin Plant Ecology*, edited by J. S. Fralish, R. P. McIntosh, and O. L. Loucks, 51–56. Madison: Wisconsin Academy of Sciences, Arts, and Letters.

Kline, V. M., and E. A. Howell. 1987. "Prairies." In *Restoration Ecology: A Synthetic Approach to Ecological Research*, edited by W. R. Jordan III, M. E. Gilpin, and J. D. Aber, 75–84. Cambridge: Cambridge University Press.

Kroger, R., S. C. Pierce, K. A. Littlejohn, M. T. Moore, and J. L. Farris. 2012. "Decreasing Nitrate-N Loads to Coastal Ecosystems with Innovative Drainage Management Strategies in Agricultural Landscapes: An Experimental Approach." *Agricultural Water Management* 103:162–66.

Kunkel, K., L. Stevens, S. Stevens, L. Sun, E. Janssen, D. Wuebbles, S. Hilberg, et al. 2013. "Regional Climate Trends and Scenarios for the U.S." *National Climate Assessment: Part 3 Climate of the Midwest*. U.S. Technical Report NESDIS 142-3. Washington, DC: U.S. Department of Commerce National Oceanic and Atmospheric Administration.

Kurtz, C. 1996. *Iowa's Wild Places: An Exploration with Carl Kurtz*. Ames: Iowa State Press.

Lack, J. B., M. J. Hamilton, J. K. Braun, M. A. Mares, and R. A. Van Den Bussche. 2013. "Comparative Phylogeography of Invasive *Rattus rattus* and *Rattus norvegicus* in the U.S. Reveals Distinct Colonization Histories and Dispersal." *Biological Invasions* 15:1067–87.

Lal, R., R. Follett, B. Stewart, and J. Kimble. 2007. "Soil Carbon Sequestration to Mitigate Climate Change and Advance Food Security." *Soil Science* 172:943–56.

Landers, R. Q., P. Christiansen, and T. Heine. 1970. "Establishment of Prairie Species in Iowa." In *Proceedings of a Symposium on Prairie and Prairie Restoration*, edited by P. Schramm, 48–49. Galesburg, IL: Knox College.

Landscape Change Research Group. 2014. *Climate Change Atlas*. Delaware, OH: USDA Forest Service Northern Research Station. Accessed February 10, 2017. http://www.fs.fed.us/nrs/atlas/.

Landwehr, K., and B. L. Rhoads. 2003. "Depositional Response of a Headwater Stream to Channelization, East Central Illinois, USA." *River Research and*

Applications 19:77–100.

Lathrop, R. C., B. M. Johnson, T. B. Johnson, M. T. Vogelsang, S. R.
Carpenter, T. R. Hrabik, J. F. Kitchell, et al. 2002. "Stocking Piscivores
to Improve Fishing and Water Clarity: A Synthesis of the Lake Mendota
Biomanipulation Project." *Freshwater Biology* 47:2410–24.

Laurance, W. F., and E. Yensen. 1991. "Predicting the Impacts of Edge Effects in
Fragmented Habitats." *Biological Conservation* 55:77–92.

Lave, R. 2009. "The Controversy over Natural Channel Design: Substantive
Explanations and Potential Avenues for Resolution." *Journal of the American
Water Resources Association* 45:1519–32.

Lavoie, C., C. Dufresne, and F. Delisle. 2005. "The Spread of Reed Canary
Grass (*Phalaris arundinacea*) in Quebec: A Spatio-Temporal Perspective."
Ecoscience 12:366–75.

Lawler, J. 2009. "Climate Change Adaptation Strategies for Resource
Management and Conservation Planning." *Annals of the New York Academy
of Sciences* 1162:79–98.

Lee, J. C., R. A. Haack, J. F. Negron, J. J. Witcosky, and S. J. Seybold. 2007.
Invasive Bark Beetles. Forest Insect and Disease Leaflet 176. Washington, DC:
U.S. Department of Agriculture Forest Service.

Lemke, A. M., J. R. Herkert, J. W. Walk, and K. D. Blodgett. 2017. "Application
of Key Ecological Attributes to Assess Early Restoration of River Floodplain
Habitats: A Case Study." *Hydrobiologia* 804:19–33.

Lemke, M. J., H. M. Hagy, K. Dungey, A. F. Casper, A. M. Lemke, T. D.
VanMiddlesworth, and A. Kent. 2017. "Echoes of a Flood Pulse: Short-Term
Effects of Record Flooding of the Illinois River on Floodplain Lakes under
Ecological Restoration." *Hydrobiologia* 804:151–75.

Lemke, M. J., and D. J. Jenkins. 2006. *LaGrange Wetland Nitrogen Dynamics:
Final Report*. Springfield: University of Illinois Springfield.

Lemke, M. J., D. Jenkins, J. Bartletti, and T. Goode. 2007. "Comparison of
Nitrogen and Bacterial Dynamics in Spunky Bottoms and LaGrange
Floodplain Wetlands." In *Spunky Bottoms: Restoration of a Big-River
Floodplain: Proceedings of the Spunky Bottoms Restoration Symposium May
2003*. Special Publication 29, edited by E. J. Heske, J. R. Herkert, K. D.
Blodgett, and A. M. Lemke, 12–14. Champaign: Illinois Natural History
Survey.

Lenhart, C. F. 2000. "The Vegetation and Hydrology of Impoundments after
Dam Removal in Southern Wisconsin." Master's thesis, University of

Wisconsin–Madison.

Lenhart, C. F., H. Peterson, and J. Nieber. 2011. "Increased Streamflow in Agricultural Watersheds of the Midwest: Implications for Management." *Watershed Science Bulletin* 2:25–31.

Leopold, A. 1933. *Game Management*. New York: Charles Scribner's Sons.

Leopold, A. 1949. *A Sand County Almanac*. New York: Oxford University Press.

Leopold, L. B., M. G. Wolman, and J. P. Miller. 1964. *Fluvial Processes in Geomophology*. New York: Dover Publications.

Lester, R. E., and A. J. Boulton. 2008. "Rehabilitating Agricultural Streams in Australia with Wood: A Review." *Environmental Management* 42:310–326.

Leuschner, W. A., J. A. Young, S. A. Walden, and F. W. Ravlin. 1996. "Potential Benefits of Slowing the Gypsy Moth's Spread." *Southern Journal of Applied Forestry* 20:65–73.

Liebhold, A., J. A. Halverson, and G. A. Elmes. 1992. "Gypsy Moth Invasion in North America: A Quantitative Analysis." *Journal of Biogeography* 19:513–20.

Lindeman, R. L. 1942. "The Trophic-Dynamic Aspect of Ecology." *Ecology* 23:399–418.

Lowenthal, D., ed. 1965. *Man and Nature by George Perkins Marsh*. Cambridge, MA: Harvard University Press.

Lubowski, R. N., A. J. Plantinga, and R. N. Stavins. 2008. "What Drives Land-Use Change in the United States? A National Analysis of Landowner Decisions." *Land Economics* 84:529–50.

Lundeen, B. 2014. *Minnesota Statewide Altered Watercourse Project*. Saint Paul: Minnesota Pollution Control Agency. Accessed February 26, 2017. http://www.pca.state.mn.us/index.php/water/water-types-and-programs/surface-water/streams-and-rivers/minnesota-statewide-altered-watercourse-project.html.

Lyons, J., and C. C. Courtney. 1990. *A Review of Fisheries Habitat Improvement Projects in Warmwater Streams, with Recommendations for Wisconsin*. Technical Bulletin No. 169. Madison: Wisconsin Department of Natural Resources.

Mahl, U. H., J. L. Tank, S. S. Roley, and R. T. Davis. 2015. "Two-Stage Ditch Floodplains Enhance N-Removal Capacity and Reduce Turbidity and Dissolved P in Agricultural Streams." *Journal of the American Water Resources Association* 51:923–40.

Makarewicz, J. C., and P. Bertram. 1991. "Evidence for the Restoration of the Lake Erie Ecosystem." *Bioscience* 41:216–23.

Maltz, M. R., and K. K. Treseder. 2015. "Sources of Inocula Influence Mycorrhizal Colonization of Plants in Restoration Projects: A Meta-Analysis." *Restoration Ecology* 23:625–34.

Matthews, J. W., and A. G. Endress. 2008. "Performance Criteria, Compliance Success, and Vegetation Development in Compensatory Mitigation Wetlands." *Environmental Management* 41:130–41.

Mattingly, R. L., E. E. Herricks, and D. M. Johnston. 1993. "Channelization and Levee Construction in Illinois: Review and Implications for Management." *Environmental Management* 17:781–95.

Mawdsley, J., R. O'Malley, and D. Ojima. 2009. "A Review of Climate-Change Adaptation Strategies for Wildlife Management and Biodiversity Conservation." *Conservation Biology* 23:1080–89.

Maynard, E., and J. Brewer. 2013. "Restoring Perennial Warm-Season Grasses as a Means of Reversing Mesophication of Oak Woodlands in Northern Mississippi." *Restoration Ecology* 21:242–49.

McClelland, M. A., G. G. Sass, T. R. Cook, K. S. Irons, N. N. Michaels, T. M. O'Hara, and C. S. Smith. 2012. "The Long-Term Illinois River Fish Population Monitoring Program." *Fisheries* 37:340–50.

McCullough, D. G., and R. J. Mercader. 2012. "Evaluation of Potential Strategies to Slow Ash Mortality (SLAM) Caused by Emerald Ash Borer (*Agrilus planipennis*): SLAM in an Urban Forest." *International Journal of Pest Management* 58:9–23.

McDonald, T., G. D. Gann, J. Jonson, and K. W. Dixon. 2016. *International Standards for the Practice of Ecological Restoration—Including Principles and Key Concepts*. Washington, DC: Society for Ecological Restoration.

McGaw, M. 2002. "The Response of Gray Dogwood (*Cornus racemosa*) to Prescribed Fire and the Effects of Invasion on Fuel Loading and Plant Community Composition at Curtis Prairie." Master's thesis, University of Wisconsin–Madison.

McIntosh, R. P. 1981. "Succession and Ecological Theory." In *Forest Succession: Concepts and Application*, edited by D. C. West, H. H. Shugart, and D. F. Botkin, 10–23. New York: Springer-Verlag.

———. 1986. *The Background of Ecology: Concept and Theory*. New York: Cambridge University Press.

———. 1995. "H. A. Gleason's 'Individualistic Concept' and Theory of Animal Communities: A Continuing Controversy." *Biological Reviews* 70:317–57.

MDA. 2015. *Golf Course Contamination from Pesticide Use*. Saint Paul: Minnesota

Department of Agriculture.

Meine, C. 1988. *Aldo Leopold: His Life and Work*. Madison: University of Wisconsin Press.

Meyer, C. K., S. G. Baer, and M. R. Whiles. 2008. "Ecosystem Recovery across a Chronosequence of Restored Wetlands in the Platte River Valley." *Ecosystems* 11:193–208.

Michaels, N. N., and G. G. Sass. 2009. *The Nature Conservancy's Emiquon Preserve: Fish and Aquatic Vegetation Monitoring Annual Report*. Technical Report 2009 (10). Champaign: Illinois Natural History Survey.

Michalak, A. M., E. J. Anderson, D. Beletsky, S. Boland, N. S. Bosch, T. B. Bridgeman, J. D. Chaffin, et al. 2013. "Record-Setting Algal Bloom in Lake Erie Caused by Agricultural and Meteorological Trends Consistent with Expected Future Conditions." *Proceedings of the National Academy of Sciences* 110:6448–52.

Middleton, B. A. 2002. "Flood Pulsing in the Regeneration and Maintenance of Species in Riverine Forested Wetlands of the Southeastern United States." In *Flood Pulsing in Wetlands: Restoring the Natural Hydrological Balance*, edited by B. A. Middleton, 223–94. New York: John Wiley and Sons.

Miles, I., W. C. Sullivan, and F. E. Kuo. 1998. "Ecological Restoration Volunteers: The Benefits of Participation." *Urban Ecosystems* 2:27–41.

Millar, C., N. Stephenson, and S. Stephens. 2007. "Climate Change and Forests of the Future: Managing in the Face of Uncertainty." *Ecological Applications* 17:2145–51.

Millennium Ecosystem Assessment. 2003. *Ecosystems and Human Well-Being: A Framework for Assessment*. Washington, DC: Island Press.

Miller, B. P., E. A. Sinclair, M. H. M. Menz, C. P. Elliott, E. Bunn, L. E. Commander, E. Dalziell, et al. 2017. "A Framework for the Practical Science Necessary to Restore Sustainable, Resilient, and Biodiverse Ecosystems." *Restoration Ecology* 25:605–17.

Miller, R. M., and J. D. Jastrow. 1986. "Influence of Soil Structure Supports Agricultural Role for Prairies, Prairie Restoration." *Restoration and Management Notes* 4:62–63.

Miller, T. P., J. R. Peterson, C. F. Lenhart, and Y. Nomura. 2012. *The Agricultural BMP Handbook* for Minnesota. Saint Paul: Minnesota Department of Agriculture.

Miller, W. 1915. *The Prairie Spirit of Landscape Gardening*. Illinois Agricultural Experiment Station Circular 184. Urbana: University of Illinois.

Milly, P., J. Betancourt, M. Falkenmark, M. Hirsch, Z. Kundzewicz, D. Lettenmaier, and R. Stouffer. 2008. "Stationarity Is Dead: Whither Water Management?" *Science* 319:573–74.

MISAC. 2015. *Minnesota Urban and Community Forest Best Management Practices for Preventing the Introduction, Establishment, and Spread of Invasive Species.* Saint Paul: Minnesota Invasive Species Advisory Council.

Mishra, V., K. Cherkauer, and S. Shukla. 2010. "Assessment of Drought Due to Historic Climate Variability and Projected Future Climate Change in the Midwestern United States." *Journal of Hydrometeorology* 11:46–68.

Mitsch, W. J. 1992. "Landscape Design and the Role of Created, Restored, and Natural Riparian Wetlands in Controlling Nonpoint Source Pollution." *Ecological Engineering* 1:27–47.

Moerke, A. H., and G. A. Lamberti. 2004. "Restoring Stream Ecosystems: Lessons from a Midwestern State." *Restoration Ecology* 12:327–34.

Mooney, H., A. Larigauderie, M. Cesario, T. Elmquist, O. Hoegh-Guldberg, S. Lavorel, G. Mace, et al. 2009. "Biodiversity, Climate Change, and Ecosystem Services." *Current Opinion in Environmental Sustainability* 1:46–54.

Moore, M., S. P. Romano, and T. Cook. 2010. "Synthesis of Upper Mississippi River System Submersed and Emergent Aquatic Vegetation: Past, Present, and Future." *Hydrobiologia* 640:103–14.

Moser, S., and J. Ekstrom. 2010. "A Framework to Diagnose Barriers to Climate Change Adaptation." *Proceedings of the National Academy of Sciences* 107:22026–31.

Moyle, P. B. 1986. "Fish Introductions into North America: Patterns and Ecological Impact." In *Ecology of Biological Invasions of North America and Hawaii*, edited by H. A. Mooney and J. A. Drake, 27–43. New York: Springer-Verlag.

MPCA. 2014. *Addressing Lakes in Watershed Restoration and Protection Strategies.* Saint Paul: Minnesota Pollution Control Agency.

Mulholland, P. J., J. L. Tank, D. M. Sanzone, J. R. Webster, W. Wollheim, B. J. Peterson, and J. L. Meyer. 2000. "Ammonium and Nitrate Uptake Lengths in a Small Forested Stream Determined by 15N Tracer and Short-Term Nutrient Enrichment Experiments." *Verhandlungen Des Internationalen Verein Limnologie* 27:1320–25.

Mulholland, S. L., S. C. Mulholland, J. R. Hamilton, T. Martin, C. Widga, and T. Lindahl. 2011. *Phase III Archaeological Data Recovery at the Christina-Pelican Site for the Lake Christina Restoration Project, Douglas and Grant Counties,*

Minnesota. Duluth, MN: Duluth Archaeology Center.

Murcia, C., J. Aronson, G. H. Kattan, D. Moreno-Mateos, K. Dixon, and D. Simberloff. 2014. "A Critique of the 'Novel Ecosystem' Concept." *Trends in Ecology and Evolution* 29:548–53.

Nassauer, J. I., M. Santelmann, and D. Scavia. 2007. *From the Corn Belt to the Gulf: Societal and Environmental Implications of Alternative Agricultural Futures*. Washington, DC: Resources for the Future.

National Golf Foundation. 2012. *U.S. Golf Industry Overview*. Jupiter, FL: National Golf Foundation.

National Invasive Species Council. 2001. *Meeting the Invasive Species Challenge: National Invasive Species Management Plan*. Washington, DC: National Invasive Species Council.

———. 2008. *2008–2012 National Invasive Species Management Plan*. Washington, DC: National Invasive Species Council.

National Research Council. 1992. *Restoration of Aquatic Ecosystems*. Washington, DC: National Academy Press.

Newbury, R. W., and M. N. Gaboury. 1993. *Stream Analysis and Fish Habitat Design: A Field Manual*. Gibsons, British Columbia: Newbury Hydraulics Ltd.

Nieber, J. L., C. Arika, C. Lenhart, M. Titov, and K. Brooks. 2011. *Evaluation of Buffer Width on Hydrologic Function, Water Quality, and Ecological Integrity of Wetlands: Final Report 2011–06*. Saint Paul: Minnesota Department of Transportation.

Nowacki, G., and M. Abrams. 2008. "The Demise of Fire and 'Mesophication' of Forests in the Eastern United States." *Bioscience* 58:123–38.

NRCS. 2000. *A Report to Congress on Aging Infrastructure*. Washington, DC: USDA Natural Resources Conservation Service.

———. 2007. "Stream Restoration Design." *Part 654 National Engineering Handbook*. Washington, DC: USDA Natural Resources Conservation Service.

———. 2017. *Web Soil Survey*. Washington, DC: USDA Natural Resources Conservation Service National Cooperative Soil Survey. Accessed June 29, 2017. https://websoilsurvey.sc.egov.usda.gov/.

Odum, E. 1971. *Fundamentals of Ecology*. Philadelphia: W. B. Saunders.

Ohio Environmental Protection Agency. 2005. *Total Maximum Daily Loads for the Euclid Creek Watershed*. Twinsburg, OH: Ohio Environmental Protection Agency Division of Surface Water.

O'Leary, C., and J. Shuey. 2003. "Ecosystem Restoration at the Landscape-Scale: Design and Implementation at the Efroymson Restoration." In *Proceedings of the 18th North American Prairie Conference: Promoting Prairie*, edited by S. Fore, 124–26. Kirksville, MO: Truman State University Press.

Olivia, M. 1955. *Living Fence Program Agreement*. Mother Theresa Hackelmeier Memorial Library Archives. Marian University, Indianapolis, IN.

Olson, K. R., and L. W. Morton. 2013. "Restoration of 2011 Flood-Damaged Birds Point–New Madrid Floodway." *Journal of Soil and Water Conservation* 68:13A–18A.

Osborne, L. L., and D. A. Kovacic. 1993. "Riparian Vegetated Buffer Strips in Water Quality Restoration and Stream Management." *Freshwater Biology* 29:243–58.

Palmer, M. A., D. A. Falk, and J. B. Zedler, eds. 2006. "Ecological Theory and Restoration Ecology." In *Foundations of Restoration Ecology*, 1–10. Washington, DC: Island Press.

Pauly, W. R. 1985. *How to Manage Small Prairie Fires*. Madison, WI: Dane County Park Commission.

Pavao-Zuckerman, M. A. 2008. "The Nature of Urban Soils and Their Role in Ecological Restoration in Cities." *Restoration Ecology* 16:642–49.

Pepoon, H. S. 1927. *An Annotated Flora of the Chicago Region*. Chicago: Chicago Academy of Sciences.

Perring, M., R. Standish, and R. Hobbs. 2013. "Incorporating Novelty and Novel Ecosystems into Restoration Planning and Practice in the 21st Century." *Ecological Processes* 2:1–8.

Perring, M. P., R. J. Standish, J. N. Price, M. D. Craig, T. E. Erickson, K. X. Ruthrof, A. S. Whiteley, et al. 2015. "Advances in Restoration Ecology: Rising to the Challenges of the Coming Decades." *Ecosphere* 6:131. doi:10.1890/es15-00121.1.

Petersen, R. C., L. B. M. Petersen, and J. Lacoursier. 1992. "A Building-Block Model for Stream Restoration." In *River Conservation and Management*, edited by P. J. Boon, P. Calow, and G. E. Petts, 293–309. Chichester, UK: John Wiley.

Peterson, B. J., W. M. Wollheim, P. J. Mulholland, J. R. Webster, J. L. Meyer, J. L. Tank, E. Martí, et al. 2001. "Control of Nitrogen Export from Watersheds by Headwater Streams." *Science* 292: 86–90.

Phelps, Q. E., S. J. Tripp, D. P. Herzog, and J. E. Garvey. 2015. "Temporary Connectivity: The Relative Benefits of Large River Floodplain Inundation in the Lower Mississippi River." *Restoration Ecology* 23:53–56.

Pickett, S. T. A., M. L. Cadenasso, J. M. Grove, C. G. Boone, P. M. Groffman, E. Irwin, S. S. Kaushal, et al. 2011b. "Urban Ecological Systems: Scientific Foundations and a Decade of Progress." *Journal of Environmental Management* 92:331–62.

Pickett, S. T. A., S. J. Meiners, and M. L. Cadenasso. 2011a. "Domain and Propositions of Succession Theory." In *The Theory of Ecology*, edited by S. M. Scheiner and M. R. Willig, 185–216. Chicago: University of Chicago Press.

Pickett, S. T. A., and T. V. Parker. 1994. "Avoiding the Old Pitfalls: Opportunities in a New Discipline." *Restoration Ecology* 2:75–79.

Pilcher, M. W., G. H. Copp, and V. Szomolai. 2004. "A Comparison of Adjacent Natural and Channelized Stretches of a Lowland River." *Biologia* 59:669–73.

Pimentel, D., R. Zuniga, and D. Morrison. 2005. "Update on the Environmental and Economic Costs Associated with Alien-Invasive Species in the United States." *Ecological Economics* 52:273–88.

Plant Conservation Alliance. 2015. *National Seed Strategy for Rehabilitation and Restoration: 2015–2020*. Washington, DC: Plant Conservation Alliance Federal Committee.

Poff, N. L., B. D. Richter, A. H. Arthington, S. E. Bunn, R. J. Naiman, E. Kendy, M. Acreman, et al. 2010. "The Ecological Limits of Hydrologic Alteration (ELOHA): A New Framework for Developing Regional Environmental Flow Standards." *Freshwater Biology* 55:147–70.

Pohl, M. M. 2002. "Bringing Down Our Dams: Trends in American Dam Removal Rationales." *Journal of American Water Resources Association* 38:1511–19.

Powell, G. E., A. D. Ward, D. E. Mecklenburg, J. Draper, and W. Word. 2007b. "Two-Stage Channel Systems: Part 2, Case Studies." *Journal of Soil and Water Conservation* 62:286–96.

Powell, G. E., A. D. Ward, D. E. Mecklenburg, and A. D. Jayakaran. 2007a. "Two-Stage Channel Systems: Part 1, A Practical Approach for Sizing Agricultural Ditches." *Journal of Soil and Water Conservation* 62:277–86.

Pryor, S., D. Scavia, C. Downer, M. Gaden, L. Iverson, R. Nordstrom, J. Patz, and G. Robertson. 2014. "Midwest." Chap. 7 in *Climate Change Impacts in the United States· The Third National Climate Assessment*, edited by J. M. Melillo, T. C. Richmon, and G. W. Yohe, 418–40. Washington, DC: U.S. Global Change Research Program.

Race, M., and M. Fonseca. 1996. "Fixing Compensatory Mitigation: What Will

It Take?" *Ecological Applications* 6:94–101.

Randall, D., R. A. Wood, S. Bony, R. Colman, T. Fichefet, J. Fyfe, V. Kattsov, et al. 2007. "Climate Models and Their Evaluation." In *Climate Change 2007: The Physical Science Basis, Contribution of Working Group I to the Fourth Assessment Report of the Intergovernmental Panel on Climate Change*, edited by S. Solomon, D. Qin, M. Manning, Z. Chen, M. Marquis, K. B. Averyt, M. Tignor, and H. L. Miller, 589–662. Cambridge: Cambridge University Press.

Real, L. A., and J. A. Brown. 1991. *Foundations of Ecology*. Classic Papers with Commentaries. Chicago: University of Chicago.

Reichard, S. H., and C. W. Hamilton. 1997. "Predicting Invasions of Woody Plants Introduced into North America." *Conservation Biology* 11:193–203.

Rhoads, B. L., and E. E. Herricks. 1996. "Naturalization of Headwater Streams in Illinois: Challenges and Possibilities." In *River Channel Restoration: Guiding Principles for Sustainable Projects*, edited by A. Brookes and F. D. Shields Jr., 331–67. Chichester, UK: John Wiley and Sons.

Rhoads, B. L., and K. D. Massey. 2012. "Flow Structure and Channel Change in a Sinuous Grass-Lined Stream within an Agricultural Drainage Ditch: Implications for Ditch Stability and Aquatic Habitat." *River Research and Applications* 28:39–51.

Rhoads, B. L., J. S. Schwartz, and S. Porter. 2003. "Stream Geomorphology, Bank Vegetation, and Three-Dimensional Habitat Hydraulics for Fish in Midwestern Agricultural Streams." *Water Resources Research* 39, no. 8: 1218. doi:10.1029/2003WR002294.

Ricciardi, A., and H. J. MacIsaac. 2000. "Recent Mass Invasion of the North American Great Lakes by Ponto-Caspian Species." *Trends in Ecology and Evolution* 15:62–65.

Ricciardi, A., and J. B. Rasmussen. 1999. "Extinction Rates of North American Freshwater Fauna." *Conservation Biology* 13:1220–22.

Rickey, M. A., and R. C. Anderson. 2004. "Effects of Nitrogen Addition on the Invasive Grass *Phragmites australis* and a Native Competitor *Spartina pectinata*." *Journal of Applied Ecology* 41:888–96.

Rinne, J. N., R. M. Hughes, and B. Calamusso. 2005. *Historical Changes in Large River Fish Assemblages of the Americas*. Symposium 45. Bethesda, MD: American Fisheries Society.

Robertson, A. I., P. Bacon, and G. Heagney. 2001. "The Responses of Floodplain Primary Production to Flood Frequency and Timing." *Journal of Applied Ecology* 38:126–36.

Rockström, J., W. Steffen, K. Noone, Å. Persson, F. S. Chapin III, E. F. Lambin, T. M. Lenton, et al. 2009. "Beyond the Boundary: A Safe Operating Space for Humanity." *Nature* 461:472–75.

Roley, S., J. L. Tank, M. L. Stephen, L. T. Johnson, J. J. Beaulieu, and J. D. Witter. 2012. "Floodplain Restoration Enhances Denitrification and Reach-Scale Nitrogen Removal in an Agricultural Stream." *Ecological Applications* 22:281–97.

Roni, P., K. Hanson, and T. Beechie. 2008. "Global Review of the Physical and Biological Effectiveness of Stream Habitat Rehabilitation Techniques." *North American Journal of Fisheries Management* 28:856–90.

Rosgen, D. 1993. "River Restoration Utilizing Natural Stability Concepts." In *Proceedings of Watershed '93: A National Conference on Watershed Management*, 783–90. Washington, DC: U.S. Environmental Protection Agency.

———. 1994. "A Classification of Natural Rivers." *Catena* 22:169–99.

———. 1996. *Applied River Morphology*. Pagosa Springs, CO: Wildland Hydrology.

———. 2007. "Rosgen Geomorphic Channel Design." In *Part 654 National Engineering Handbook*, 11.1–11.76. Washington, DC: USDA Natural Resources Conservation Service.

Ross, J. 1984. "Managing the Public Rangelands: 50 Years Since the Taylor Grazing Act." *Rangelands* 6:147–51.

Sachse, N. D. 1974. *A Thousand Ages: The University of Wisconsin Arboretum*. Madison: University of Wisconsin Press.

Sauer, L. 1998. *The Once and Future Forest: A Guide to Forest Restoration Strategies*. Washington, DC: Island Press.

Saunders, D., R. Hobbs, and C. Margules. 1991. "Biological Consequences of Ecosystem Fragmentation: A Review." *Conservation Biology* 5:18–32.

Schaller, J. L., T. V. Royer, M. B. David, and J. L. Tank. 2004. "Denitrification Associated with Plants and Sediments in an Agricultural Stream." *Journal of the North American Benthological Society* 23:667–76.

Scheffer, M. 2004. *Ecology of Shallow Lakes*. Norwell, MA: Kluwer Academic Publishers.

Scheffer, M., S. H. Hosper, M. L. Meijer, B. Moss, and E. Jeppesen. 1993. "Alternative Equilibria in Shallow Lakes." *Trends in Ecology and Evolution* 8:275–79.

Scheffer, M., and E. H. van Nes. 2007. "Shallow Lakes Theory Revisited: Various

Alternative Regimes Driven by Climate, Nutrients, Depth, and Lake Size."
Hydrobiologia 584:455–66.

Scheller, R. M., and D. J. Mladenoff. 2008. "Simulated Effects of Climate
Change, Fragmentation, and Inter-Specific Competition on Tree Species
Migration in Northern Wisconsin, USA." *Climate Research* 36:191–202.

Schindler, D. W. 2006. "Recent Advances in the Understanding and
Management of Eutrophication." *Limnology and Oceanography* 51:356–63.

Schlosser, I. J. 1987. "A Conceptual Framework for Fish Communities in Small
Warmwater Streams." In *Community and Evolutionary Ecology of North
American Stream Fishes*, edited by W. J. Matthews and D. C. Heins, 17–24.
Norman: Oklahoma University Press.

Schottler, S. P., J. Ulrich, P. Belmont, R. Moore, J. W. Lauer, D. R. Engstrom, and
J. E. Almendinger. 2013. "Twentieth Century Agricultural Drainage Creates
More Erosive Rivers." *Hydrological Processes* 28:1951–61.

Schumm, S. A. 1979. "Geomorphic Thresholds: The Concept and Its
Applications." *Transactions of the Institute of British Geographers* 4:485–515.

———. 1981. "Evolution and Response of the Fluvial System: Sedimentologic
Implications." In *Recent and Ancient Nonmarine Depositional Environments:
Models for Exploration*, SEPM Special Publication No. 31, edited by
F. G. Ethridge and R. M. Flores, 19–29. Tulsa, OK: Society of Economic
Paleontologists and Mineralogists.

Schummer, M. L., J. Palframan, E. McNaughton, T. Barney, and S. A. Petrie.
2012. "Comparisons of Bird, Aquatic Macroinvertebrate, and Plant
Communities among Dredged Ponds and Natural Wetland Habitats at Long
Point, Lake Erie, Ontario." *Wetlands* 32:945–53.

Schwartz, J. S., and E. E. Herricks. 2005. "Fish Use of Stage-Specific Fluvial
Habitats as Refuge Patches during a Flood in a Low-Gradient Illinois
Stream." *Canadian Journal of Fisheries and Aquatic Sciences* 62:1540–52.

Sears, P. B. 1925. "The Natural Vegetation of Ohio." *The Ohio Journal of Science*
25:139–49.

Secchi, S., and B. A. Babcock. 2007. "Impact of High Corn Prices on
Conservation Reserve Program Acreage." *Iowa Ag Review* 13:4–5, 7.

SER. 2004. *The SER International Primer on Ecological Restoration*. Tucson, AZ:
Society for Ecological Restoration.

Service, R. F. 2011. "Will Busting Dams Boost Salmon?" *Science* 334:888–92.

Shafroth, P. B., J. M. Friedman, G. T. Auble, M. L. Scott, and J. H. Braatne. 2002.
"Potential Responses of Riparian Vegetation to Dam Removal." *Bioscience*

52:703–12.

Shankman, D., and T. B. Pugh. 1992. "Discharge Response to Channelization of a Coastal Plain Stream." *Wetlands* 12:157–62.

Shapiro, J., V. Lamarra, and M. Lynch. 1975. "Biomanipulation: An Ecosystem Approach to Lake Restoration." In *Water Quality Management through Biological Control*, edited by P. L. Brezonik and J. L. Fox, 85–96. Gainesville: University of Florida.

Shapiro, J., and D. I. Wright. 1984. "Lake Restoration by Biomanipulation: Round Lake, Minnesota, the First Two Years." *Freshwater Biology* 14:371–83.

Sheley, R., and J. Petroff. 1999. *Biology and Management of Noxious Rangeland Weeds.* Corvallis: Oregon State University Press.

Shelford, V. E. 1917. "The Ideals and Aims of the Ecological Society of America." *Bulletin of the Ecological Society of America* 1:1–8.

Shields, F. D., Jr., S. S. Knight, and C. M. Cooper. 1998. "Rehabilitation of Aquatic Habitats in Warmwater Streams Damaged by Channel Incision in Mississippi." *Hydrobiologia* 382:63–86.

Shuey, J. 2013. "Habitat Re-creation (Ecological Restoration) as a Strategy for Conserving Insect Communities in Highly Fragmented Landscapes." *Insects* 4:761–80.

Sietman, B. E. 2007. "Freshwater Mussels of the Minnesota River Valley Counties." In *Native Plant Communities and Rare Species of the Minnesota River Valley Counties*, 5.32–5.39. Saint Paul: Minnesota Department of Natural Resources.

Simon, A., S. J. Bennett, and J. M. Castro. 2011. *Stream Restoration in Dynamic Fluvial Systems: Scientific Approaches, Analyses, and Tools.* Geophysical Monograph Series, Volume 194. Washington, DC: American Geophysical Union.

Simmons, B. L., R. A. Hallett, N. F. Sonti, D. S. N. Auyeung, and J. W. T. Lu. 2016. "Long-Term Outcomes of Forest Restoration in an Urban Park." *Restoration Ecology* 24:109–18.

Simpson, T. B. 2008. "The Dechannelization of Nippersink Creek: Learning about Native Illinois Streams through Restoration." *Ecological Restoration* 26:350–56.

Simpson, T. W., L. A. Martinelli, A. N. Sharpley, and R. W. Howarth. 2009. "Impact of Ethanol Production on Nutrient Cycles and Water Quality: The United States and Brazil as Case Studies." In *Biofuels: Environmental Consequences and Interactions with Changing Land Use*, edited by R. W.

Howarth and S. Bringezu, 153–67. Ithaca, NY: Cornell University.

Skidmore, P. B., F. D. Shields Jr., M. W. Doyle, and D. E. Miller. 2001. "A Categorization of Approaches to Natural Channel Design." In *Proceedings of the 2001 Wetlands Engineering and River Restoration Conference*. Reston, VA: American Society of Civil Engineers.

Smiley, P. C., Jr., and B. J. Allred. 2011. "Differences in Aquatic Communities between Wetlands Created by an Agricultural Water Recycling System." *Wetlands Ecology and Management* 19:495–505.

Smiley, P. C., Jr., and R. B. Gillespie. 2010. "Influence of Physical Habitat and Agricultural Contaminants on Fishes within Agricultural Drainage Ditches." In *Agricultural Drainage Ditches: Mitigation Wetlands for the 21st Century*, edited by M. T. Moore and R. Kroger, 37–73. Kerula, IN: Research Signpost.

Smiley, P. C., Jr., K. W. King, and N. R. Fausey. 2011. "Influence of Herbaceous Riparian Buffers on Physical Habitat, Water Chemistry, and Stream Communities within Channelized Agricultural Headwater Streams." *Ecological Engineering* 37:1314–23.

Smiley, P. C., Jr., and K. R. Rumora. 2015. "Influence of Planting Grass Filter Strips on the Structure and Function of Riparian Habitats of Agricultural Headwater Streams." *Riparian Ecology and Conservation* 2:58–71.

Smith, D. 1995. "Integrated Roadside Vegetation Management: The Iowa Model." In *Proceedings of Fifth International Symposium on Environmental Concerns in Rights-of-Way Management*, edited by G. C. Doucet and M. G. Seguin, 191–93. Montreal, CA: Hydro-Quebec.

———. 2014. "Prairie Restoration: Bridging the Past and the Future." In *Proceedings of the 23rd North American Prairie Conference*, edited by C. Jacques, 62–69. Brookings: South Dakota State University.

Smith, D., D. Williams, G. Houseal, and K. Henderson. 2010. *The Tallgrass Prairie Center Guide to Prairie Restoration in the Upper Midwest*. Iowa City: University of Iowa Press.

Smith, E. P., D. R. Orvos, and J. Cairns Jr. 1991. "Impact Assessment Using the Before-After-Control-Impact (BACI) Model: Concerns and Comments." *Canadian Journal of Fisheries and Aquatic Sciences* 50:627–37.

Snow, K. 2002. "Working with the Horticulture Industry to Limit Invasive Species Introductions." *Conservation in Practice* 3:33–38.

Snyder, T. A., III. 2004. "A Spatial Analysis of Grassland Species Richness in Curtis Prairie." Master's thesis, University of Wisconsin–Madison.

Solomon, L. E., R. M. Pendleton, A. F. Casper, N. T. Grider, and R. B.

Hilsabeck. 2014. *Changes in the Fish Community at The Nature Conservancy's Merwin Preserve at Spunky Bottoms*. Technical Report 2014 (31). Champaign: Illinois Natural History Survey.

Sparks, R. E. 1995. "Need for Ecosystem Management of Large Rivers and Their Floodplains." *Bioscience* 45:168–82.

Sparks, R. E., K. D. Blodgett, A. F. Casper, H. M. Hagy, M. J. Lemke, L. Machado-Velho, and L. Cleide-Rodrigues. 2017. "Why Experiment with Success? Opportunities and Risks in Applying Assessment and Adaptive Management to the Emiquon Floodplain Restoration Project." *Hydrobiologia* 804:177–200.

Sparks, R. E., J. Nelson, and Y. Yin. 1998. "Naturalization of the Flood Regime in Regulated Rivers: The Case of the Upper Mississippi River." *Bioscience* 48: 706–20.

Sperry, T. M. 1984. "Analysis of the University of Wisconsin–Madison Prairie Restoration Project." In *Proceedings of the 8th North American Prairie Conference*, edited by R. Brewer, 140–47. Kalamazoo: Western Michigan University.

Stafford, J. D., M. M. Horath, A. P. Yetter, R. V. Smith, and C. S. Hine. 2010. "Historical and Contemporary Characteristics and Waterfowl Use of Illinois River Valley Wetlands." *Wetlands* 30:565–76.

Staudinger, M., T. Morelli, and A. Bryan. 2015. *Integrating Climate Change into Northeast and Midwest State Wildlife Action Plans*. Amherst, MA: Department of Interior Northeast Climate Science Center.

Stein, B., A. Staudt, M. S. Cross, N. S. Dubois, C. Enquist, R. Griffis, L. Hansen, et al. 2013. "Preparing for and Managing Change: Climate Adaptation for Biodiversity and Ecosystems." *Frontiers in Ecology and the Environment* 11:502–10.

Stevens, W. K. 1995. *Miracle under the Oaks*. New York: Pocket Books.

Stewart-Oaten, A., W. W. Murdoch, and K. R. Parker. 1986. "Environmental Impact Assessment: Pseudoreplication in Time." *Ecology* 67:929–40.

Stradling, D., and R. Stradling. 2008. "Perceptions of the Burning River: Deindustrialization and Cleveland's Cuyahoga River." *Environmental History* 13:515–35.

Strock, J. S., C. J. Dell, and J. P. Schmidt. 2007. "Managing Natural Processes in Drainage Ditches for Nonpoint Source Nitrogen Control." *Journal of Soil and Water Conservation* 62:188–96.

Suding, K. N. 2011. "Toward an Era of Restoration in Ecology: Successes,

Failures, and Opportunities Ahead." *Annual Review Ecology, Evolution, and Systematics* 42:465–87.

Suding, K. N., and K. L. Gross. 2006. "The Dynamic Nature of Ecological Systems: Multiple States and Restoration Trajectories." In *Foundations of Restoration Ecology*, edited by D. Falk, M. Palmer, and J. Zedler, 190–209. Washington, DC: Island Press.

Swink, F. A., and G. Wilhem. 1979. *Plants of the Chicago Region*. Lisle, IL: Morton Arboretum.

Tallamy, D. W. 2004. "Do Alien Plants Reduce Insect Biomass?" *Conservation Biology* 18:1689–92.

Tester, J., and M. Keirstead. 1995. *Minnesota's Natural Heritage: An Ecological Perspective*. Minneapolis: University of Minnesota Press.

Theiling, C. H. 1995. "Habitat Rehabilitation on the Upper Mississippi River." *Regulated Rivers: Research and Management* 11:227–38.

Thompson, D. M., and G. N. Stull. 2002. "The Development and Historic Use of Habitat Structures in Channel Restoration in the United States: The Grand Experiment in Fisheries Management." *Geographie Physique et Quaternaire* 56:45–60.

Thompson, W. H. 1989. *Transportation in Iowa: A Historical Summary*. Ames: Iowa Department of Transportation.

Tibbetts, J. 1997. "Exotic Invasion." *Environmental Health Perspectives* 105:590–93.

Tilman, D., and J. A. Downing. 1994. "Biodiversity and Stability in Grasslands." *Nature* 367:363–65.

Tilman, D., P. B. Reich, and J. M. Knops. 2006. "Biodiversity and Ecosystem Stability in a Decade-Long Grassland Experiment." *Nature* 441:629–32.

Tobin, G. A. 1995. "The Levee Love Affair: A Stormy Relationship?" *Journal of the American Water Resources Association* 31:359–67.

Tockner, K., F. Malard, and J. V. Ward. 2000. "An Extension of the Flood Pulse Concept." *Hydrological Processes* 14:2861–83.

Tockner, K., and J. A. Stanford. 2002. "Riverine Flood Plains: Present State and Future Trends." *Environmental Conservation* 29:308–30.

Trabucco, A., R. Zomer, D. Bossio, O. Van Straaten, and L. Verchot. 2008. "Climate Change Mitigation through Afforestation/Reforestation: A Global Analysis of Hydrologic Impacts with Four Case Studies." *Agriculture, Ecosystems and Environment* 126:81–97.

Transeau, E. 1935. "The Prairie Peninsula." *Ecology* 16:423–37.

Tungesvick, K. 2003. *Floral Inventory of the Marian College EcoLab.* Muncie, IN: Spence Restoration Nursery.

Urban, M. A., and B. L. Rhoads. 2003. "Catastrophic Human-Induced Change in Stream-Channel Planform and Geometry in an Agricultural Watershed, Illinois, USA." *Annals of the Association of American Geographers* 93:783–96.

USACE. 1991. *Lake Chautauqua Rehabilitation and Enhancement Definite Project Report.* Rock Island, IL: U.S. Army Corps of Engineers, Rock Island District.

———. 2001. *Peoria Lake Habitat Rehabilitation and Enhancement Project: Initial Performance and Evaluation Report.* Rock Island, IL: U.S. Army Corps of Engineers, Rock Island District.

———. 2004. *Banner Marsh Habitat and Rehabilitation Program Performance and Evaluation Report.* Rock Island, IL: U.S. Army Corps of Engineers, Rock Island District.

U.S. Census Bureau. 2016. *Metropolitan and Micropolitan Statistical Area Datasets: 2010–2015: April 1, 2010 to July 1, 2015.* Accessed June 21, 2017. https://www.census.gov/data/datasets/2015/demo/popest/total-metro-and-micro-statistical-areas.html.

Van Cleef, J. S. 1885. "How to Restore Our Trout Streams." *Transactions of the American Fisheries Society* 14:53–55.

van der Valk, A. G., and C. B. Davis. 1978. "The Role of Seed Banks in the Vegetation Dynamics of Prairie Glacial Marshes." *Ecology* 59:322–35.

Van Diggelen, R., A. P. Grootjans, and J. A. Harris. 2001. "Ecological Restoration: State of the Art or State of the Science?" *Restoration Ecology* 9:115–18.

Van Kleunen, M., E. Weber, and M. Fischer. 2010. "A Meta-Analysis of Trait Differences between Invasive and Non-Invasive Species." *Ecology Letters* 123:235–45.

VanMiddlesworth, T. D., N. N. McClelland, G. G. Sass, A. F. Casper, T. W. Spier, and M. J. Lemke. 2017a. "Fish Community Succession and Biomanipulation to Control Two Common Aquatic Ecosystem Stressors during a Large-Scale Floodplain Lake Restoration." *Hydrobiologia* 804:73–88.

VanMiddlesworth, T. D., G. G. Sass, B. A. Ray, T. W. Spier, J. Lyons, N. N. McClelland, and A. F. Casper. 2017b. "Food Habits and Relative Abundances of Native Piscivores: Implications for Controlling Common Carp." *Hydrobiologia* 804:89–101.

Verry, E. S. 2004. "Land Fragmentation and Impacts to Streams and Fish in the Central and Upper Midwest." In *A Century of Forest and Wildland Watershed*

Lessons, edited by G. G. Ice and J. D. Stednick, 129–54. Bethesda, MD: Society of American Foresters.

Vogt, C., and D. Thompson. 1973. *Marian College Wetlands Ecological Laboratory Site Plan*. Minneapolis: Minnesota Environmental Sciences Foundation, Inc.

Vought, L. B. M., and J. O. Lacoursiere. 2010. "Restoration of Streams in the Agricultural Landscape." In *Restoration of Lakes, Streams, Floodplains, and Bogs in Europe: Principles and Case Studies*. Vol. 1, *Wetlands, Ecology, Conservation and Management*, edited by M. Eiseltova, 225–42. Dordrecht, NL: Springer.

Waller, D. M., K. L. Amatangelo, S. Johnson, and D. A. Rogers. 2012. "Plant Community and Resurvey Data from the Wisconsin Plant Ecology Laboratory." *Biodiversity and Ecology* 4:255–64.

Walsh, C. J., A. H. Roy, J. W. Feminella, P. D. Cottingham, P. M. Groffman, and R. P. Morgan II. 2005. "The Urban Stream Syndrome: Current Knowledge and the Search for a Cure." *Journal of the North American Benthological Society* 24:706–23.

Walter, T., M. Dosskey, M. Khanna, J. Miller, M. Tomer, and J. Wiens. 2007. "The Science of Targeting within Landscapes and Watersheds to Improve Conservation Effectiveness." In *Managing Agricultural Landscapes for Environmental Quality*, edited by M. Schnepf and C. Cox, 63–89. Ankey, IA: Soil and Water Conservation Society.

Ward, J. V., K. Tockner, U. Uehlinger, and F. Malard. 2001. "Understanding Natural Patterns and Processes in River Corridors as the Basis for Effective River Restoration." *Regulated Rivers: Research and Management* 17:311–23.

Waters, T. F. 1995. *Sediment in Streams: Sources, Biological Effects, and Control*. Bethesda, MD: American Fisheries Society.

Weaver, J. 1954. *North American Prairie*. Lincoln, NE: Johnson Publishing Company.

Wegner, M., P. Zedler, B. Herrick, and J. Zedler. 2008. *Curtis Prairie: 75-Year-Old Restoration Research Site*. Arboretum Leaflets No. 16. Madison: University of Wisconsin–Madison Arboretum.

Westphal, L. M., P. H. Gobster, and M. Gross. 2010. "Models for Renaturing Brownfield Areas." In *Restoration and History: The Search for a Usable Environmental Past*, edited by M. Hall, 208–17. New York: Routledge.

Wheaton, J. M., S. E. Darby, D. A. Sear, and J. A. Milne. 2006. "Does Scientific Conjecture Accurately Describe Restoration Practice? Insight from an International River Restoration Survey." *Area* 38: 138–42.

Whisenant, S. G. 1999. *Repairing Damaged Wildlands: A Process-Oriented, Landscape-Scale Approach*. Cambridge: Cambridge University Press.

White, R. J. 1996. "Growth and Development of North American Stream Habitat Management for Fish." Supplement 1, *Canadian Journal of Fisheries and Aquatic Sciences* 53:342–63.

Wilcock, D., and F. Wilcock. 1995. "Modeling the Hydrological Impacts of Channelization on Streamflow Characteristics in a Northern Ireland Catchment." In *Modeling and Management of Sustainable Basin-Scale Water Resource Systems*, edited by S. P. Simonovic, Z. Kundzewicz, D. Rosbjerg, and K. Takeuchi, 41–49. Oxfordshire, UK: International Association of Hydrological Sciences.

Williams, D. B. 2002. "The Sparrow War." *American History* 37:38–43.

Williams, J. E., C. A. Wood, and M. P. Dombeck. 1997. "Understanding Watershed-Scale Restoration." In *Watershed Restoration: Principles and Practices*, edited by J. E. Williams, C. A. Wood, and M. P. Dombeck, 1–13. Bethesda, MD: American Fisheries Society.

Winter, T. 1989. "Hydrologic Studies of Wetlands in the Northern Prairie." In *Northern Prairie Wetlands*, edited by A. van der Valk, 16–54. Ames: Iowa State University Press.

Wright, J., and G. Sands. 2001. *Planning an Agricultural Subsurface Drainage System*. Agricultural Drainage Series Publication # 07685. Saint Paul: University of Minnesota Extension.

Yarber, B. 1959. "Attractive Campus, Wishing Bridge, Await Wanderers." *Marian College Phoenix* (Indianapolis, IN): 3.

Zedler, J. B. 2000a. "Progress in Wetland Restoration Ecology." *Trends in Ecology and Evolution* 15:402–7.

———. 2000b. *Handbook for Restoring Tidal Wetlands*. Boca Raton, FL: CRC Press.

———. 2007. "Success: An Unclear, Subjective Descriptor of Restoration Outcomes." *Ecological Restoration* 25:162–68.

Zomer, R., A. Trabucco, D. Bossio, and L. Verchot. 2008. "Climate Change Mitigation: A Spatial Analysis of Global Land Suitability for Clean Development Mechanism Afforestation and Reforestation." *Agriculture, Ecosystems, and Environment* 126:67–80.